Anelasticity in the Earth

Geodynamics Series

Inter-Union Commission on Geodynamics
Editorial Board
A. L. Hales, Chairman
R. D. Russell, Secretary
O. L. Anderson
F. M. Delany
C. L. Drake
J. Sutton

American Geophysical Union/Geological Society of America
Editorial Board
Kees Dejong
C. L. Drake
D. H. Eckhardt
E. Irving
W. I. Rose
Rob Van der Voo

The Final Reports of the International Geodynamics Program sponsored by the Inter-Union Commission on Geodynamics.

Anelasticity in the Earth

Edited by F. D. Stacey
M. S. Paterson
A. Nicholas

Geodynamics Series
Volume 4

American Geophysical Union
Washington, D.C.
Geological Society of America
Boulder, Colorado
1981

Final Report (Part A) of Working Group 5, Properties and Processes of the Earth's Interior coordinated by O. L. Anderson on behalf of the Bureau of the Inter-Union Commission on Geodynamics.

American Geophysical Union, 2000 Florida Avenue, N.W.
Washington, D.C. 20009

Geological Society of America, 3300 Penrose Place, P.O. Box 9140
Boulder, Colorado 80301

Library of Congress Cataloging in Publication Data

Main entry under title:

Anelasticity in the earth.

(Geodynamics series; v. 4)
Includes bibliographies.
1. Earth--Mantle--Addresses, essays, lectures.
2. Elasticity--Addresses, essays, lectures. I. Stacey, F. D. (Frank D.) II. Paterson, M. S. III. Nicolas, A. (Adolphe), 1936- IV. Series.
QE509.A63 1981 551.1'16 81-10944
ISBN 0-87590-505-6 AACR2

Printed in the United States of America

CONTENTS

FOREWORD

After a decade of intense and productive scientific cooperation between geologists, geophysicists and geochemists the International Geodynamics Program formally ended on July 31, 1980. The scientific accomplishments of the program are represented in more than seventy scientific reports and in this series of Final Report volumes.

The concept of the Geodynamics Program, as a natural successor to the Upper Mantle Project, developed during 1970 and 1971. The International Union of Geological Sciences (IUGS) and the International Union of Geodesy and Geophysics (IUGG) then sought support for the new program from the International Council of Scientific Unions (ICSU). As a result the Inter-Union Commission on Geodynamics was established by ICSU to manage the International Geodynamics Program.

The governing body of the Inter-Union Commission on Geodynamics was a Bureau of seven members, three appointed by IUGG, three by IUGS and one jointly by the two Unions. The President was appointed by ICSU and a Secretary-General by the Bureau from among its members. The scientific work of the Program was coordinated by the Commission, composed of the Chairmen of the Working Groups and the representatives of the national committees for the International Geodynamics Program. Both the Bureau and the Commission met annually, often in association with the Assembly of one of the Unions, or one of the constituent Associations of the Unions.

Initially the Secretariat of the Commission was in Paris with support from France through BRGM, and later in Vancouver with support from Canada through DEMR and NRC.

The scientific work of the Program was coordinated by ten Working Groups.

WG 1 Geodynamics of the Western Pacific-Indonesian Region

WG 2 Geodynamics of the Eastern Pacific Region, Caribbean and Scotia Arcs

WG 3 Geodynamics of the Alpine-Himalayan Region, West

WG 4 Geodynamics of Continental and Oceanic Rifts

WG 5 Properties and Processes of the Earth's Interior

WG 6 Geodynamics of the Alpine-Himalayan Region, East

WG 7 Geodynamics of Plate Interiors

WG 8 Geodynamics of Seismically Inactive Margins

WG 9 History and Interaction of Tectonic, Metamorphic and Magmatic Processes

WG 10 Global Syntheses and Paleoreconstruction

These Working Groups held discussion meetings and sponsored symposia. The papers given at the symposia were published in a series of Scientific Reports. The scientific studies were all organized and financed at the national level by national committees even when multinational programs were involved. It is to the national committees, and to those who participated in the studies organized by those committees, that the success of the Program must be attributed.

Financial support for the symposia and the meetings of the Commission was provided by subventions from IUGG, IUGS, UNESCO and ICSU.

Information on the activities of the Commission and its Working Groups is available in a series of 17 publications: Geodynamics Reports, 1-8, edited by F. Delany, published by BRGM; Geodynamics Highlights, 1-4, edited by F. Delany, published by BRGM; and Geodynamics International, 13-17, edited by R. D. Russell. Geodynamics International was published by World Data Center A for Solid Earth Geophysics, Boulder, Colorado 80308, USA. Copies of these publications, which contain lists of the Scientific Reports, may be obtained from WDC A. In some cases only microfiche copies are now available.

This volume is one of a series of Final Reports summarizing the work of the Commission. The Final Report volumes, organized by the Working Groups, represent in part a statement of what has been accomplished during the Program and in part an analysis of problems still to be solved. This volume, one of two from W. G. 5 (Chairman, O.L. Anderson) was edited by F.D. Stacey, M.S. Paterson and A. Nicolas.

At the end of the Geodynamics Program it is clear that the kinematics of the major plate movements during the past 200 million years is well understood, but there is much less understanding of the dynamics of the processes which cause these movements.

Perhaps the best measure of the success of the Program is the enthusiasm with which the Unions and national committees have joined in the establishment of a successor program to be known as: Dynamics and evolution of the lithosphere: The framework for earth resources and the reduction of the hazards.

To all of those who have contributed their time so generously to the Geodynamics Program we tender our thanks.

C. L. Drake, President ICG, 1971-1975

A. L. Hales, President ICG, 1975-1980

Members of Working Group 5:

O.L. Anderson
G. Barta
W.S. Fyfe
S. Akimoto
V. Babuska
E. Boschi
A.H. Cook
C. Froidevaux
K. Fuchs
I.G. Gass
P. Grew
D.G. Kautzleben
J.G. Negi
A. Nicolas
L. O. Nicolaysen
R.J. O'Connell
M.G. Rochester
C.H. Scholz
N.V. Sobolev
F.D. Stacey
H. Stiller
L.P. Vinnik
K. Yagi

PREFACE

The fact that the materials in the solid parts of the Earth depart from perfectly elastic behaviour is now central to fundamental geophysical studies. Slow deformation (creep) by the processes of mantle convection has been widely recognised for about 25 years; it has been realised for much longer that seismic waves are attenuated but more recently there has developed a general if vague awareness of a probable correlation between high attenuation and low resistance to creep and in the last few years seismic attenuation has received much greater attention because it implies a first order dispersion of body waves (frequency dependence of elasticity).

Two bodies, the International Geodynamics Commission (IGC) through its Working Group 5 on Properties and Processes in the Earth's Interior, and the International Association for Seismology and Physics of the Earth's Interior (IASPEI) independently recognised that the time was ripe for a coordinated symposium on these problems. By good fortune they approached the same proposed conveners with suggestions for such a symposium to be held at the IUGG General Assembly in Canberra in December 1979. The result was a meeting supported by both bodies that brought together experts on the problems of attenuation and creep. This volume comprises selected papers from the proceedings and indicates the present state of these subjects, with emphasis on recent progress.

It is now clear that at room temperature and pressure and at sufficiently low strain amplitudes (less than about 10^{-6}) damping of elastic waves is linear in the sense that the principle of superposition is valid. It is important that measurements at high temperatures and pressures be made to determine the strain range over which damping is linear under these conditions. Assuming it to be similar, then except very close to the source regions, the strains of seismic waves are well within the linear range and we can apply linear viscoelastic theory to seismic wave propagation. In the special case that has usually been assumed, that the anelastic damping parameter Q is independent of frequency, the theory is relatively straight-forward, but the concensus of seismologists now is that Q shows a general increase with increasing frequency, being at least twice as great for high frequency body waves as at free oscillation periods. This not only complicates the viscoelastic theory but makes the problem of inverting seismological data much more difficult, especially as there is no reason to believe that the frequency dependence of Q is the same at all depths, any more than is its value. However, the observation of the frequency dependence of Q is important to elucidation of the mechanism of attenuation and this is of interest also in relating to creep properties.

Current opinion generally favours dislocation motion as the principal mechanism of creep deformation in the mantle, with climb being possibly the rate-controlling step. However, the empirical activation energies measured for creep do not agree with the activation energies for diffusion of individual atomic species, so that the rate-controlling process is evidently not simple atomic diffusion within the grains. Also, experiments so far have necessarily concentrated on rather simple materials, generally monomineralic (olivine) and often single crystals, and we must exercise caution in applying parameters derived from these experiments directly to the mantle. With respect to attenuation, data on mantle materials under mantle conditions are not available and extrapolation of laboratory observations on the basis of established Solid State Physics is necessary. This is another area requiring more attention, but preliminary conclusions indicate the importance of crystalline defects, especially dislocations. Thus crystal dislocations may be responsible for both the plasticity and anelasticity of the mantle. Also compositional variations, relative movement at grain boundaries, and the presence of water can have important influences on both, but detailed study of these factors remains for the future.

We expect the phenomena of anelasticity and creep in the Earth to be active research topics for many years. Hopefully this report volume will provide a useful stepping stone to progress.

F.D. Stacey
Department of Physics, University of Queensland, Brisbane 4067, Australia

M.S. Paterson
School of Earth Sciences, Australian National University, Canberra, Australia

A. Nicolas
Laboratoire de Tectonophysique, Universite de Nantes, Nantes, France

THE PHYSICS OF CREEP AND ATTENUATION IN THE MANTLE

Don L. Anderson and J. Bernard Minster

Seismological Laboratory, California Institute of Technology, Pasadena, California 91125

Abstract. Dislocations contribute to both seismic wave attenuation and steady-state creep in the mantle. The two phenomena involve quite different strains and characteristic times but they can both be understood with simple dislocation models. The most satisfactory model for creep involves the glide of dislocations across subgrains and rate limited by the climb of jogged dislocations in the walls of a polygonized network. The jog formation energy contributes to the apparent activation energy for creep, making it substantially larger than the activation energy for self-diffusion. The theory leads to either a σ^2 or σ^3 law for creep rate, depending on the length of the dislocations relative to the equilibrium spacing of thermal jogs.

Attenuation in the mantle at seismic frequencies is probably caused by the glide of dislocations in the subgrains. Kink and impurity drag can both contribute to the glide time constant. The kink-formation, or Peierls barrier, model for dislocation glide appears to be a low-temperature, high-frequency mechanism most appropriate for pure systems. A small amount of impurity drag brings the dislocation glide characteristic time into the seismic band at upper-mantle temperatures.

The attenuation and creep behavior of the mantle are related through the dislocation structure. Discussion of the various possible mechanisms is facilitated by casting them and the geophysical data in terms of a pre-exponential characteristic time and an activation energy. The relaxation strength is an additional parameter that can be used to identify the attenuation mechanism. Mobile dislocations in subgrains, rather than cell walls, have the appropriate characteristics to explain the damping of seismic waves in the upper mantle. The grain boundary peak may be responsible for attenuation in the lithosphere.

I. Introduction

The purpose of this paper is to investigate the role of dislocations in both creep and attenuation. The high temperature creep of most crystalline solids is controlled by dislocation climb or glide plus climb [Weertmann, 1968]. For a wide variety of metallic, ionic, and polar crystals the activation energies for creep and self-diffusion are esentially identical, indicating that the motions of dislocations are rate-limited by self-diffusion. Furthermore, in the stress range of about 10^{-2} to 10^{-5} G, where G is the shear modulus, the stress exponent of the creep rate is about 3, which is consistent with dislocation climb or the glide of jogged dislocations. Both of these mechanisms of steady-state creep require self-diffusion. The third-power creep law is satisfied by olivine in the stress range of several hundred bars to several kilobars [Kohlstedt and Goetze, 1974; Goetze and Kohlstedt, 1973; Kohlstedt *et al.*, 1976].

The laboratory creep law for olivine is commonly assumed to hold for the much lower stresses and strain rates appropriate for the mantle, and this has been the basis for the calculation of many viscosity profiles for the mantle. It has also been assumed that the activation energy of creep, 125 kcal/mole, represents the diffusional activation energy for the slowest moving species. It has recently been found, however, that the activation energies for diffusion of oxygen and silicon in olivine are only about 90 kcal/mole [Reddy *et al.*, 1980; Jaoul *et al.*, 1979]. In addition, at low stresses the stress exponent may be closer to two than to three.

The mechanism of attenuation in crystalline solids also appears to be an activated process. Dislocation damping is a high-temperature mechanism that is rate-limited by the diffusion of point defects. The glide of dislocations does not necessarily require self-diffusion and can therefore operate on a much shorter time scale than climb.

II. a. Activated Processes

Thermally activated relaxation processes, including those that are controlled by diffusion, are functions of a characteristic relaxation time that can be written

$$\tau = \tau_0 \exp (E^*/RT) \tag{1}$$

where τ_0, the pre-exponential, is the relaxation time at infinite temperature and E^* is the effective activation energy. For simple point defect mechanisms τ_0^{-1} is close to the Debye frequency, $\sim 10^{13}$ sec^{-1}. For mechanisms involving dislocations or dislocation-point defect interactions, τ_0 depends on other parameters such as dislocation lengths, Burgers vectors, kink and jog separations, Peierls stress, interstitial concentrations, etc.

The activation energy, depending on the mechanism, is a composite of activation energies relating to self-diffusion, kink or jog formation, Peierls energy, bonding energy of point defects to the dislocation, core-diffusion, and so on. E^* for creep of metals is often just the self-diffusion activation energy, E_{SD}, which indicates that the other contributions are small. This is not necessarily the case for silicates.

Non-elastic processes in geophysics are commonly expressed in terms of viscosity for long-term phenomena, and Q for oscillatory and short-term phenomena. It is convenient to express these in terms of the characteristic time. For creep

$$\tau = \sigma/G\dot{\epsilon} = \eta/G \tag{2}$$

where σ, G, $\dot{\epsilon}$, and η are the deviatoric stress, rigidity, strain rate, and viscosity, respectively. Thus, the characteristic time can be simply computed from creep theories and experiments.

Relaxation theories of attenuation can be described by

$$Q^{-1} = 2Q_0^{-1}\, \omega\tau/(1 + \omega^2\tau^2) \tag{3}$$

where ω is the frequency, Q^{-1} is the standard measure of attenuation, and Q_0^{-1} is the peak value of attenuation at $\omega\tau = 1$. In the mantle, where Q is low, we infer that the characteristic relaxation time is close to $1/\omega$, the reciprocal of the measurement frequency. High values of Q indicate that $\omega \gg 1/\tau$ or $\omega \ll 1/\tau$. In these cases $Q \sim \omega$ or ω^{-1}. When Q is slowly varying we are probably in the midst of an absorption band which is conveniently explained as the result of a distribution of relaxation times [Liu *et al.,* 1977; Anderson *et al.,* 1977; Minster, 1980].

Since Q is small in the higher temperature regions of the mantle, it is likely that at seismic periods the mantle is on the low-temperature side of an absorption band. Thus

$$Q^{-1} = 2Q_0^{-1}/\omega\tau = (2Q_0^{-1}/\omega\tau_0)\exp(-E^*/RT) \qquad (4)$$

in regions of rapidly varying attenuation. In this paper we use the seismic data to estimate both the relaxation time in the upper mantle and its variation with depth. Details of the attenuation in the midst of an absorption band and the effect of a spectrum of relaxation times is discussed in a companion paper (Minster and Anderson, 1981). High Temperature Background (HTB) is a dominant mechanism of attenuation in crystalline solids at high temperature and low frequency. This usually satisfies a $Q \sim \omega$ or $Q \sim \omega^{\alpha}$ relation. The latter has been interpreted by Anderson and Minster (1978) in terms of a distribution of relaxation times. The former is valid for all relaxation mechanisms when the measurement frequency is sufficiently high. In the following $\hat{\tau}_0$ will be used for the pre-exponential creep, or Maxwell, relaxation time and τ_0 will be used for the attenuation, or Q, characteristic time.

II. b. Activation Energies for Climb and Glide

The creep of ductile materials such as metals is rate limited by self-diffusion. This is the case for either pure diffusional creep or dislocation climb. In these circumstances the activation energy for creep is the same as that for self-diffusion. In olivine the activation energy for creep is appreciably higher than that for self-diffusion of O^{2-} or Si^{4+}. This suggests that creep may be controlled by kink or jog formation [Gueguen, 1979].

The activation energy for creep when dislocation climb is due to nucleation and lateral drift of jogs is $(1/2)\ (E^*_{SD} + E^*_{CD} + 2E^*_j)$ or $(1/2)\ (E^*_{SD} + E^*_{CD} + 4E^*_j)$ depending on whether the dislocations are longer or shorter than the equilibrium jog spacing. Here E^*_{SD}, E^*_{CD} and E^*_j are the activation energies for self-diffusion, core-diffusion and jog formation, respectively [Hirth and Lothe, 1968].

E^*_{SD} is 90 kcal/mole for both oxygen and silicon diffusion [Reddy *et al.,* 1980, Jaoul *et al.,* 1979]. The jog formation energy is somewhat greater than kink formation energy [Hirth and Lothe, 1968]. The activation energy controlling high-temperature deformation of olivine has been estimated by Goetze and Kohlstedt [1973] to be 135 kcal/mole from the climb of dislocation loops in olivine and 122-128 kcal/mole by Ashby and Verrall [1978] and Goetze [1978] from creep data on olivine.

In the absence of interstitial defects near the dislocation the activation energy for glide is E^*_k if the kink density is high and $2E^*_k$ when the dislocation length is shorter than the equilibrium distance between kinks. E^*_k is the kink formation energy which Stocker and Ashby [1973] estimate to be 26 kcal/mole. The activation energy for glide is therefore 52 kcal/mole if the kink spacing is large or if double kink formation is required for glide. If point defects must diffuse with the dislocation line then their diffusivity may be rate limiting. The activation energy for Mg-Fe interdiffusion in olivine is about 58 kcal/mole for 0.9 mole fraction Mg, decreasing to 50 kcal/mole for pure forsterite [Misener, 1974]. Dislocation glide in olivine will therefore have an activation energy of the order of 50-60 kcal/mole unless there are slower moving species in the vicinity of the dislocation. Olivine in contact with pyroxene is likely to have excess silicon interstitials as a dominant point defect [Stocker, 1978]. The activation energy in this case may be due to drag of silicon interstitials, about 90 kcal/mole. There is also a small term representing the binding energy of interstitials to the dislocation. This is ordinarily a few kcal/mole [Van Bueren, 1961] but data are lacking for silicates. The kink and jog formation energies in silicates are also highly uncertain and the above estimates for olivine are approximate.

The activation energy for dislocation core diffusion should be similar to that for surface diffusion. This, for metals, is usually about 1/2 to 2/3 of the lattice diffusion value. If this rule applies to olivine the core diffusion activation energy would be about 50 to 60 kcal/mole. The hot-pressing experiments of Schwenn and Goetze [1978] yield a value of 85 kcal/mole, close to that for lattice diffusion of O^{2-} and Si^{4+}. The dislocation mobility experiments of Goetze and Kohlstedt [1974] also suggests that the activation energies for lattice and core diffusion are similar. The actual values determined in their experiments may be interpreted as the sum of the diffusion and jog formation energies. A high value for the core and surface activation energies is expected from the heat of vaporization, which is much higher for silicates than for metals.

If we adopt $E_{SD} = E_{CD} = 90$ kcal/mole and $E_{creep} = 125$ kcal/mole then the implied jog formation energy is 35 kcal/mole. This is much greater than is typical of metals. This is also expected since the kink and jog formation energies are proportional to Gb^3 which is much greater for silicates than for metals.

Because of the high values for core diffusion and jog formation we expect relatively high values for the creep activation energy for silicates and a non-correspondance with the self-diffusional activation energy, except for Coble and Nabarro-Herring creep which will dominate for very small grain sizes and low-stresses.

A summary of the physical properties of olivine is assembled in Table 1. These will be used in subsequent calculations.

II. c. Geophysical Constraints on the Activation Parameters

Some of the activation parameters for creep and attenuation can be estimated directly from the geophysical data. The characteristic time scales of relaxation processes in the mantle depend on temperature, through the activation energy, and the pressure, through the activation volume. Stress has an indirect effect on relaxation times since it controls such parameters as dislocation density and subgrain size. The characteristic frequency of atomic processes and processes involving dislocation motions are typically 10^{-13} to 10^{-10} seconds. The characteristic relaxation time for creep in the upper mantle, the ratio of

Table 1

Material Properties for Olivine

Property	Symbol	Value	Units
Burger's vector	b	6×10^{-8}	cm
Oxygen ion volume	Ω	1×10^{-23}	cm^3
Shear modulus	G	8×10^{11}	dy/cm^2
Silicon diffusivity(1)			
pre-exponential	D_{Si}	1.5×10^{-6}	cm^2/sec
activation energy	E_{Si}	90	kcal/mole
Oxygen diffusivity(2)			
pre-exponential	D_{ox}	3.5×10^{-3}	cm^2/sec
activation energy	E_{ox}	89	kcal/mole
Mg-Fe diffusivity(3)			
pre-exponential	D_{Mg}	3.4×10^{-3}	cm^2/sec
activation energy	E_{Mg}	47	kcal/mole
Subgrain size:			
Dislocation length	K′	15	—
King energy(4)	E_k	26	kcal/mole
Jog energy(5)	E_j	35	kcal/mole

(1) Jaoul et al., 1979; (2) Reddy et al., 1980; (3) Misener, 1974; (4) Stocker and Ashby, 1973; (5) From creep data

viscosity to rigidity, is about 10^{10} seconds. For temperatures of 1500-1600 K, appropriate for the upper mantle, the required activation energy is 145-170 kcal/mole. This can be compared with the activation energy for creep of olivine 125-165 kcal/mole [Goetze, 1978, Jaoul *et al.*, 1979] and similar values interred for the climb of dislocations in olivine [Goetze, 1978, Kohlstedt *et al.*, 1976].

Judging from the attenuation of surface waves [Anderson and Archambeau, 1964, Anderson and Hart, 1978] the characteristic time controlling and attenuation of seismic waves in the upper mantle is of order of 10^2 seconds. This implies an activation energy in the range 90-100 kcal/mole for the above τ_0. This is close to the activation energy found for diffusion of O^{2-} and Si^{4+} in olivine [Reddy *et al.*, 1980; Jaoul *et al.*, 1979]. A higher value for τ_0 gives a smaller value for E*.

The trade-off between τ_0 and E* is shown in Figure 1. The curves to the right cover a range of upper mantle temperatures and viscosities. The curves to the left represent the seismic band. Theories of creep and attenuation predict τ_0 – E* pairs. Those which fall in the areas shown are geophysically plausible mechanisms. Low values of τ_0 require high values for E*. As we will show, the characteristic pre-exponential times are not the same for creep and attenuation.

It is not simple to estimate the pre-exponential characteristic time appropriate for attenuation in the mantle. High-temperature background damping processes in metals give τ_0 in the range of 10^{-12} – 10^{-14} seconds [Nowick and Berry, 1972]. Lower temperature peaks give characteristic times one or two orders of magnitude higher [Woirgard, 1976]. Theoretically, the times scale as the diffusivity which is typically three orders of magnitude lower in silicates than in metals. On the other hand, subgrain sizes and dislocation lengths in the mantle can be expected to be one or two orders of magnitude greater than typical laboratory grain sizes in metals. Since $\tau_0 \sim \ell^2/D$ the characteristic time for the mantle may therefore be of the order of 10^{-6} to 10^{-8} seconds. There is very limited data on silicates. For sintered forsterite with grain size of 5×10^{-4} cm Jackson [1969] obtained 4×10^{-13} sec for a relatively low-temperature relaxation peak. His temperatures were not high enough to resolve the HTB peak but one can calculate from his data that $\tau_0 \leqslant 10^{-7}$ seconds. The subgrain size in olivine at 10 bars, a reasonable mantle stress, is of the order of 10^{-1} cm [Durham *et al.*, 1977]. The mantle relaxation times can therefore be estimated from the forsterite data as about 10^{-8} sec. This implies an activation energy of 58-74 kcal/mole if the upper mantle seismic absorption band is centered at 100 seconds.

Although τ is proportional to ℓ^2 for most attenuation and creep mechanisms there are other factors which affect the characteristic times. Steady-state creep is controlled by self-diffusion of the slowest moving species. This affects both D_0 and E*. E* for creep, which also includes the energy of jog formation, is therefore greater than E* for attenuation which need not involve self-diffusion. For a polygonized network the climb of dislocations in cell walls is rate-limiting but the characteristic time for creep involves both the motion of the long mobile dislocations in the cells and the shorter dislocations in the walls. For attenuation, the separate population of dislocations would give rise to widely separated internal friction peaks. A wide separation of characteristic times can therefore be obtained from a single dislocated solid.

Although there is a large uncertainty in the τ_0 appropriate for attenuation in the mantle the range of inferred activation energies is consistent with diffusion control. The lower end of the range is consistent with activation energies found at high temperatures, low frequencies and low strains in Al_2O_3 and Mg_2SiO_4 [Jackson, 1969]. These in turn are comparable to those appropriate for interstitial drag (Mg-Fe) or double-kink formation.

II. d. Estimates of τ_0 and E* for Olivine

The characteristic relaxation time for creep of olivine can be obtained from high-temperature laboratory experiments. The relatively low stress, <2 kbar, creep data satisfies

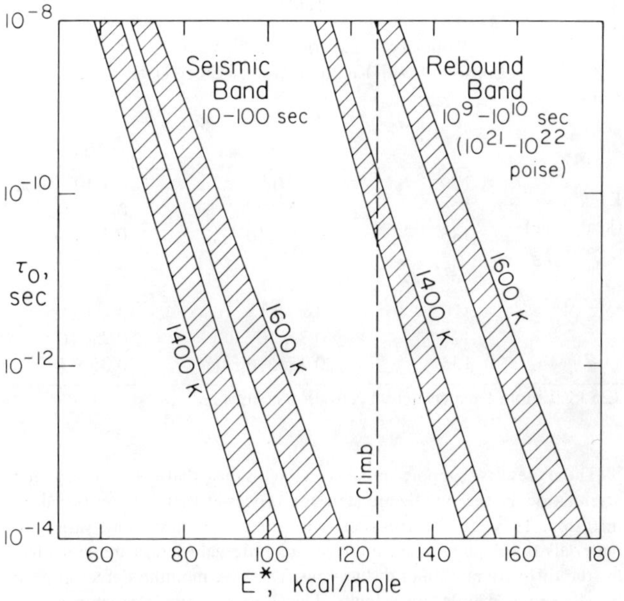

Fig. 1. Pre-exponential τ_0 vs. activation energy combinations that satisfy the seismic data (left) and rebound data (right) for a range of upper mantle temperatures.

$$\dot{\epsilon} = A\sigma^n \exp(-E/RT) \quad (5)$$

Using the Kohlstedt-Goetze data with n = 3 and E = 125 kcal/mole and the relation

$$\hat{\tau}_0 = \sigma G/\dot{\epsilon}_0 \quad (6)$$

we obtain $\hat{\tau}_0 = 3 \times 10^{-9}$ to 3×10^{-11} sec for the stress range 1 - 10 bars. The low-stress data has a σ^2 trend. In this case $\hat{\tau}_0$ is 10^{-9} to 10^{-10} sec. A summary of $\hat{\tau}_0$ values consistent with the olivine creep data is given in Table 2. A range of activation energies is given.

From the rigidity and viscosity the relaxation time of the upper mantle is $10^{10\pm1}$ seconds. Using an activation energy of 125 kcal/mole and a range of upper mantle temperatures from 1100 to 1300°C the inferred value for $\hat{\tau}_0$ is 10^{-6} to 10^{-10} seconds. This is consistent with the laboratory creep data for differential stresses in the range of a fraction of a bar to about 5 bars. The mantle $\hat{\tau}_0$ values also overlap the lower values for $\hat{\tau}_0$ determined from dislocation mechanisms rate limited by silicon diffusion.

The effect of pressure can be estimated from the observation that both viscosity and Q do not vary by more than 2 orders of magnitude in the mantle [Anderson and Hart, 1978, Peltier, 1979]. This constrains the activation volume to be between 4 and 9 cm^3/mole. Since the activation volume of oxygen, the largest major ion in the mantle, is about 11 cm^3/mole and decreases with pressure and since the effective activation volume depends on all diffusing ions for coupled diffusion, the result from the mantle is consistent with diffusional control for both dislocation creep and attenuation.

A crude estimate of the activation energy controlling attenuation can be obtained from the observation that Q can vary by about an order of magnitude over relatively short distances both laterally and vertically in the upper mantle. Using 200°C as a reasonable difference in temperature the inferred activation energy is about 52 kcal/mole: This, in turn, requires a τ_0 of about 10^{-5} – 10^{-6} seconds for the attenuation mechanism.

Table 2

Summary of Experimental $\hat{\tau}_0$ Values Consistent with Kohlstedt-Goetze Creep Data

	σ^3–law	Stress	
		1 bar	10 bar
E*	120	3×10^{-8} sec.	3×10^{-10} sec.
(kcal/mole)	125	0.7×10^{-8}	0.7×10^{-10}
	135	0.35×10^{-9}	0.35×10^{-11}
	σ^2–law		
	120	3×10^{-11}	3×10^{-12}
	125	0.7×10^{-11}	0.7×10^{-12}
	135	0.35×10^{-12}	0.35×10^{-13}

125 kcal/mole is experimental activation energy

These results support previous conclusions that both creep and attenuation in the mantle are activated processes that are controlled by diffusion. They are also consistent with numerous studies on a variety of materials that high temperature creep and internal friction are controlled by the diffusive motion of dislocations. The large modulus defect implied by the upper mantle low-velocity, low-Q zone is also consistent with a dislocation relaxation mechanism [Gueguen and Mercier, 1972; Anderson and Minster, 1980]. Point defect internal friction peaks are generally relatively sharp, occur at low temperatures and do not exhibit the low Q's and large velocity dispersion that characterize the mantle and the high temperature internal friction peaks.

III. Calculation of Creep of Olivine

The climb velocity of a dislocation, when both self-diffusion and jog formation are important is [Hirth and Lothe, 1968]

$$v = (4\pi D_{SD} \sigma b^2/kT) \exp(\Delta E - 2E_j)/2RT \qquad (7)$$

We use $D_{SD} = D_{Si} = (D_{Si})_0 \exp(-E_{Si}/RT)$ since Si^{4+} appears to be the slowest moving species. We assume that the material contains cells or subgrains of diameter L. The creep equation is [Gittus, 1976]

$$\dot{\epsilon} = K^6 \sigma^3 D\Omega^{1/3}/G^2 kT \qquad (8)$$

where $\sigma = Gb/L$ has been taken as the effective stress operating on the mobile dislocations of length L. The parameter K is the ratio of cell diameter to the average dislocation length, including mobile and cell-wall dislocations. This parameter has not been measured in olivine but the ratio of average mobile dislocation length to cell diameter is about 15 [Durham *et al.,* 1977]. This means that the Gittus parameter $K > 15$. Gittus also assumes that each subgrain contains only one mobile dislocation. Equation (8) still applies if there are more dislocations per subgrain but the K parameter would have a different value. The strain rate increases with an increase in mobile dislocation density but decreases with a decrease in length of the dislocations. K is therefore a slowly varying function of the mobile dislocation density in the subgrains. ΔE in equation (7) is $E_{SD} - E_{CD}$.

The calculated creep curves are given in Fig. 2 using D_0 for silicon and K of 29 and 15. A K of 29 fits the data. There is a direct trade-off between D_0 and K. If oxygen proves to be the slowest moving species then, with current estimates of its diffusivity, the compatible K is 8.

The subgrain structure of stressed olivine, a characteristic it shares with other crystalline materials, also affects the attenuation. If we assume that only the free dislocations contribute to the high temperature attenuation, their lengths are

$$L = K(Gb/\sigma) \qquad (9)$$

The relaxation time then depends inversely on stress squared and, with K = 10, the characteristic times are 10^2 greater than they would be for a uniform Frank network. Note that the characteristic time for creep is proportional to K^{-6}, i.e. very much shorter than the case for a uniform network.

Steady-state creep data for olivine for stresses between several hundred bars and several kilobars are well fit by a dislocation climb model where the strain rate is proportional to the third power of the stress [Goetze, 1978]. At lower stresses the data deviate from this relationship and approach a stress-squared dependency. The stress-squared dependency at low stresses is suggested by data of Kohlstedt and Goetze [1974], Durham *et al.,* [1979] and Berckhemer *et al.,* [1979].

The dislocation climb velocity for the case where the equilibrium jog spacing is greater than the average dislocation length is [Hirth and Lothe, 1968].

$$V = \frac{4\pi \ell D_{SD} \Omega b \sigma}{a^2 b kT \ell n(z/b)} \exp(\Delta E^* - 4E_j^*)/2RT \qquad (10)$$

where z is the mean free path of a core vacancy along the dislocation line (~1.4a), a is the height of a thermal jog at high temperature, ℓ is the dislocation length, D_{SD} is the pre-exponential self-diffusion coefficient (approximately equal to the diffusivity of the slowest moving species), Ω is the atomic volume, b is the Burgers vector, σ is the stress, ΔE is the difference in activation energy between self-diffusion and core diffusion. Setting $a \sim b \sim \Omega^{1/3}$ and writing $\dot{\epsilon} = \rho V b$, we have

$$\dot{\epsilon} \sim 4\pi D_{SD} b\sigma^2/kTG \qquad (11)$$

We have used the relation $Gb/\ell = \sigma$ where G is the rigidity. Equation (11) has the required stress dependency and an effective activation energy which is substantially greater than the self-diffusion activation energy.

IV. a. Attenuation

The actual physical mechanism of attenuation in the mantle is uncertain but it is likely to be a relaxation process involving a distribution of relaxation times (e.g., Anderson *et al.,* 1977, Minster, 1980). Many of

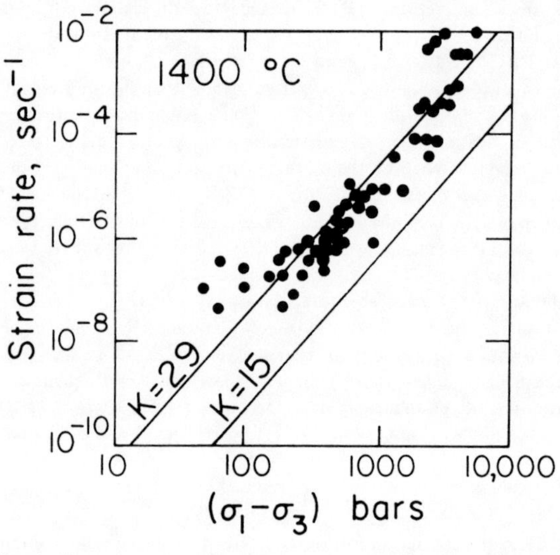

Fig. 2. Creep in polygonized olivine for several values of K. data from Kohlstedt et al. (1976).

the attenuation mechanisms that have been identified in solids occur at relatively low temperatures and high frequencies and can therefore be eliminated from consideration. These include point defect and dislocation resonance mechanisms which typically give absorption peaks at kilohertz and megahertz frequencies at temperatures below about half the melting point. The so-called grain boundary and cold-work peak and the "high temperature background" occur at lower frequencies and higher temperatures. These mechanisms involve the stress induced diffusion of dislocations. The Bordoni peak occurs at relatively low temperature in metals but may be a higher temperature peak in silicates.

Even in the laboratory it is often difficult to identify the mechanism of a given absorption peak. The effects of amplitude, frequency, temperature, irradiation, annealing, deformation and impurity content must be studied before the mechanism can be identified with certainty. This information is not available for the mantle or even for the silicates which may be components of the mantle. Nevertheless, there is some information which helps constrain the possible mechanism of attenuation in the mantle.

1. The frequency dependence of Q is weak over most of the seismic band. At frequencies greater than about 1 Hz Q appears to increase linearly with frequency [Kanamori and Anderson, 1977, Minster, 1978, Sipkin and Jordan, 1979]. This is consistent with the behavior expected on the low-temperature side of a relaxation band. A weak frequency dependence is best accomplished by invoking a distribution of relaxation times. A distribution of dislocation lengths, grain sizes and activation energies may be involved.

2. Although it has not been specifically studied, there has been no evidence brought forward to suggest that seismic attenuation is amplitude or stress dependent. Laboratory measurements of attenuation are independent of amplitude at strains less than 10^{-6}. Strains associated with seismic waves are generally much less than this.

3. The radial and lateral variations of Q are our best clues to the effects of temperature and pressure. The lower Q regions of the mantle seem to be in those areas where the temperatures are highest. This suggests that most of the upper mantle is on the low-temperature side of an absorption band or in the band itself. At a depth of 100 km the temperature of the continental lithosphere is about 200 K less than under oceans. Q is roughly 7 times larger under continents. This implies an activation energy of about 50 kcal/mole.

4. The variation of Q with depth in the mantle covers a range of less than two orders of magnitude. This means that the effects of temperature and pressure are relatively modest or that they tend to compensate each other.

5. Losses in shear are more important than losses in compression. This is consistent with stress induced motion of defects rather than a thermo-elastic mechanism or other mechanisms involving bulk dissipation.

IV. b. Dislocation Damping

The general expression for the characteristic time of stress induced diffusion of a dislocation line is

$$\tau = (kTL^2/D_0Gb^3)\exp(E^*/RT) \tag{12}$$

In the bowed string approximation for climb L is the dislocation length, D_0 is the pre-exponential self-diffusion coefficient and E* is the activation energy for self-diffusion. If the climb is rate-limited by core diffusion (CD) then [Woirgard, 1976]

$$L^2 = \ell^4/\lambda b \tag{13}$$

where ℓ is the dislocation length, $\lambda = b\exp(E_j/RT)$ is the jog separation and D_0 and E* refer to core diffusion. The effective activation energy is therefore the sum of the core diffusion and jog formation energies. If $D_0(CD) = 10^3\,D_0(SD)$ and $\ell = 10^3$ b the τ_0 for core-diffusion is 10^3 times that for self-diffusion. At modest temperature, however, $\tau(CD) < \tau(SD)$ if $E_{CD} < E_{SD}$. In general, relaxation peaks with high activation energies occur at higher temperature and lower frequency than those with lower activation energies. The presence of point defects, kinks or jogs along the dislocation line changes the characteristic time.

An estimate of the relaxation time for climb can be obtained from the values in Table 1. We first assume that the dislocation length is the same as the subgrain size, which in turn is related to the tectonic stress (Durham *et al.,* 1977).

$$L = 15\,Gb/\sigma \tag{14}$$

Using D_{Si} and σ between 1 and 10 bars we obtain τ_0 of $1 - 10^{-2}$ seconds. Neither the self-diffusion nor the core-diffusion characteristic times can be brought into the upper mantle seismic band with the appropriate activation energy (Table 1 and Figure 1). The actual mobile dislocation lengths in olivine are about 15 times smaller than the subgrain size. This reduces τ_0 to 4×10^{-3} to 4×10^{-5} seconds. With an activation energy of 90 kcal/mole the relaxation time at upper mantle temperatures is 10^8 to 10^6 seconds, still well outside the seismic band but straddling the Chandler period. Dislocation climb in subgrains is therefore unlikely to contribute to seismic attenuation but may contribute to damping of the Chandler wobble.

The dislocation lengths in cell walls are an order of magnitude smaller than in the cells. This makes a further reduction in τ_0 to about 10^{-5} to 10^{-7} sec. This gives relaxation times just outside the seismic band at mantle temperatures and E* = 90 kcal/mole but the relaxation strength is very small because of the limited bow-out possible for small dislocations.

Therefore, it appears that dislocation climb can be ruled out as a mechanism for attenuation in the mantle. We therefore turn our attention to dislocation glide.

In general dislocations glide much faster than they climb and the relaxation time is therefore much reduced. Glide is rate limited by lattice (Peierls) stresses or by point defects.

For a gliding dislocation, rate limited by the diffusion of interstitial defects [Schoeck, 1963]

$$L^2 = C_i\ell^2 \tag{15}$$

where C_i is the concentration of interstitials along the dislocation line. At sufficiently high temperature

$$C_i = C_D\exp(-E_B/RT) \tag{16}$$

and C_D is the bulk concentration of interstitial point defects or impurity atoms and E_B is the binding energy of the point defects to the dislocation line. The diffusivity is that appropriate for the diffusion of the impurity or interstitial ions.

Using $D_0 = 10^{-3}$ cm²/sec and $C_D \sim C_i = 10^{-3}$ we obtain τ_0 of 10^{-6} to 10^{-8} seconds for typical dislocation lengths appropriate for mantle stresses of 1 to 10 bars. These, combined with activation energies appropriate for diffusion of cations in silicates give relaxation times in the seismic band at upper mantle temperatures. The impurity content refers to that in the subgrains. Most impurities in the mantle are probably at grain boundaries and therefore the grain interiors are relatively pure. The above estimate for C_D is therefore not unreasonable.

IV. c. Relaxation Strength

The relaxation strength, or maximum value for Q^{-1}, is given by

$$2Q^{-1} = (1/6\sqrt{5})\rho_m\ell^2 \tag{17}$$

for a collection of randomly oriented gliding dislocations (Minster and Anderson, 1980). In the subgrains $\rho_m = 1/\ell^2$ for a rectangular grid of dislocations and the relaxation strength is about 7.5% or a Q of about 26. The same relationship holds for a Frank network in the cell walls but because of the small volume in the walls the average relaxation strength is very small for the wall contribution to the attenuation.

The relaxation strength is proportional to the area swept out by the gliding dislocations and gives the difference between the high-frequency or low-temperature modulus and the relaxed modulus. Gueguen and Mercier (1972) and Anderson and Minster (1980) pointed out that the velocity and attenuation in the upper mantle low-velocity zone could be explained by dislocation relaxation.

IV. d. Examples of Dislocation Mechanisms of Attenuation

The Köster peak occurs in cold-worked metals containing interstitial impurities, some of which have concentrated along the dislocations. It occurs at high temperature and low frequency and usually gives a τ_o of 10^{-13} to 10^{-14} seconds for metals with O, N or H interstitials. It is believed to be caused by an interaction between dislocations and interstitial point defects in the neighborhood of the dislocation [Schoeck, 1963]. With reasonable choices of parameters this mechanism can explain attenuation in the mantle.

The grain boundary peak (GBP) occurs in metals at about one-half the melting temperature at 1 Hz. Actually, a series of peaks is often observed between about 0.3 and 0.6 T_m. The activation energies range from about 0.5–1.0 of the self-diffusion activation energy and are higher for the high temperature peaks. An increase in solute concentration increases the magnitude of the higher temperature peak [Nowick and Berry, 1972]. These peaks are all superimposed on an even stronger high-temperature background which dominates at very high temperature. In synthetic forsterite a strong peak occurs at about 0.25 – 0.5 Hz at 1000°C [Jackson, 1968] with an activation energy of 57 kcal/mole. At upper mantle temperatures this absorption band shifts to about 10^{-2} seconds and therefore would not be important at seismic frequencies. In addition the peak is much less pronounced in natural olivine and the HTB gives greater absorption than the GBP at temperatures greater than about 1500°C. The GBP may be responsible for attenuation in the lithosphere.

The Bordoni Peak was first identified in deformed ("cold worked") metals at low temperature. It is unlikely to be important at mantle temperatures and seismic frequencies but it would be useful to know its properties for mantle minerals since it contains information about the Peierls energy.

The Hirth and Lothe [1968] theory for the characteristic relaxation time gives

$$\tau = \frac{b^2 kT}{2E_k D_k} \exp(2E_k/RT) \tag{18}$$

With nominal values for olivine the relaxation time at mantle temperatures is 10^{-7} seconds. Thus, this is a high frequency, low temperature mechanism.

The controlling activation energy is $2E_k$ or about 52 kcal/mole for olivine. This combination of τ_o and E^* make it unlikely that the Peierls' energy alone controls attenuation in the mantle. It may be responsible for the "grain boundary" peak in olivine for which Jackson [1968] obtained τ_o of 10^{-13} sec and E^* of 57 kcal/mole. The value for E_k, however, is uncertain.

V. Discussion

Values of $\hat{\tau}_o$ of order 10^{-7} to 10^{-12} seconds for the upper mantle are implied by relaxation times of 10^9 – 10^{10} seconds at temperatures of 1400-1600 K and an activation energy of 125 ± 5 kcal/mole. These values are consistent with both laboratory creep data and creep in a polygonized network model of olivine if the stresses are less than about 10 bars. This applies to both the σ^3 and σ^2 laws. If kilobar level stresses existed in the upper mantle then the inferred viscosity and relaxation times would be at least 6 orders of magnitude lower than observed. On the other hand, there is no contradiction with kilobar level stresses being maintained in the lithosphere for 10^6 years if temperatures are less than about 900 K. Kilobar level stresses also shift the absorption band to much higher frequencies.

The same dislocations contribute to phenomena with quite different time scales. The climb of jogged dislocations is rate-limited by self-diffusion and jog-nucleation. In silicates, and other materials with high Peierls energy, the activation energy for creep can be appreciably greater than for self-diffusion. At finite temperature the creep rate will therefore be slower, and the characteristic time longer, than for materials rate-limited by self-diffusion alone. The effective activation energy for creep is either E_j or $2E_j$ greater than for self-diffusion, depending on whether the dislocation length is smaller or larger than the equilibrium spacing of thermal jogs. These two situations lead to a σ^3 or σ^2 creep law. If the activation energy for core-diffusion is less than for self-diffusion the effective activation energy is reduced by one-half the difference of the two. This is a relatively small effect and appears to be negligible for olivine.

On the other hand, the glide of dislocations is controlled by a much smaller activation energy and is therefore much faster than climb at finite temperature. This is the case whether glide is controlled by kink nucleation or diffusion of interstitials. Both creep and attenuation depend on dislocation length and therefore tectonic stress. They are both also exponentially dependent on temperature although the controlling activation energies are different. In principle, the viscosity of the mantle can be estimated from the Q.

VI. Directions for Future Research

There is now abundant data on the high-temperature creep of olivine. The various creep theories can be tested with this data and relatively confident extrapolations can be made to lower strain rates. There is no inconsistency between the laboratory and geophysical data. The major uncertainty is the stress dependence and activation energy at low stresses and the microstructure at low stresses. The Gittus theory satisfies the laboratory and geophysical data. The assumption that there is only one mobile dislocation per grain and that it traverses the grain by glide alone may need modification. The theory also assumes that cell-walls of all orientation behave the same. Nevertheless, the Gittus theory seems closer to reality than theories involving climb alone in Frank networks.

The calculation of relaxation times and relaxation strengths for glide in the subgrains is straight forward and appears capable of explaining the seismic attenuation data. The laboratory data for checking the theory is basically non-existent for silicates. Low frequency, small strain experiments on deformed and annealed single and polycrystals at a variety of temperatures is required. The effect of impurity content must also be studied in both olivine and peridotite. In principle, seismic data can be used to infer temperature, dislocation density, stress and subgrain impurity content. This paper has hopefully made the case for a new generation of geophysical experiments.

Acknowledgments

This research was supported by the Earth Sciences Section, National Science Foundation Grant No. EAR77-14675, and the National Aeronautics and Space Administration Grant No. NSG-7610. Contribution Number 3411, Division of Geological and Planetary Sciences, California Institute of Technology, Pasadena, California 91125.

References

Anderson, D. L. and C. Archambeau, The anelasticity of the Earth, *J. Geophys. Res.. 69,* 2071-2084, 1964.

Anderson, D. L., H. Kanamori, R. Hart and H. Liu, The Earth as a seismic absorption band, *Science, 196,* 1104-1106, 1977.

Anderson, D. L. and B. Minster, The frequency dependence of Q in the Earth and implications for mantle rheology and Chandler Wobble, *Geophys. J. R. Astron. Soc. 58,* 431-440, 1980.

Anderson, D. L. and R. Hart, Q of the Earth, *J. Geophys. Res., 83,* 5869-5882, 1978.

Anderson, D. L., and B. Minster, Seismic velocity, attenuation and rheology of the upper mantle, in *Source Mechanism and Earthquake Prediction,* Coulomb vol. (ed. C. Allegre) p. 13-22, Centre National de la Recherche Scientifique, Paris, 1980.

Ashby, M. and R. Verrall, Micromechanisms of flow and fracture and their relevance to the rheology of the upper mantle, *Phil. Trans. R. Soc. Land A. 288,* 59-95, 1977.

Berckhemer, H., F. Auer and J. Drisler, High-temperature anelasticity and elasticity of mantle peridotite, *Phys. Earth Planet. Int. 20,* 48-59, 1979.

Celli, V., M. Kabler, T. Ninomiya and R. Thomson, Theory of dislocation mobility in semiconductors, *Phys. Rev., 131,* 58-72, 1963.

Durham, W., C. Goetze and B. Blake, Plastic flow of oriented single crystals of olivine 2. observations and interpretations of the dislocation structures, *J. Geophys. Res., 82,* 5755-5770, 1977.

Gittus, J. H., Theoretical equation for steady-state dislocation creep effects of jog-drag and cell formation, *Phil. Mag., 34,* 401-411, 1976.

Goetze, C. and D. Kohlstedt, Laboratory study of dislocation climb and diffusion in olivine, *J. Geophys. Res., 78,* 5961-5971, 1973.

Goetze, C., The mechanisms of creep in olivine, *Phil. Trans. R. Soc. Lond. A. 288,* 99-119, 1978.

Gueguen, Y., High temperature olivine creep: evidence for control by edge dislocations, *Geophys. Res. Lett., 6,* 357-360, 1979.

Gueguen, Y., and J. M. Mercier, High attenuation and the low-velocity zone, *Phys. Earth Planet. Int., 7,* 39-46, 1973.

Hirth, J. and J. Lothe, Theory of dislocations, McGraw-Hill, New York, p. 780, 1968.

Jackson, D., Grain boundary relaxation and the attenuation of seismic waves, Ph.D. Thesis, M.I.T., Cambridge, Mass., 1969.

Jaoul, O., C. Froidevaux and M. Poumellec, Atomic diffusion of ^{18}O and ^{30}Si in forsterite: implications for the high temperature creep mechanisms *ICG abstracts,* Internat. Un. of Geodesy and Geophysics, XVII, General Assembly, Canberra, Australia, 1979.

Kanamori, H. and D. L. Anderson, Importance of physical dispersion in surface-wave and free-oscillation problems; a review, *Rev. Geophys. Space Phys., 15,* 105-112, 1977.

Kohlstedt, D. and C. Goetze, Low-stress high-temperature creep in olivine single crystals, *J. Geophys. Res., 79,* 2045-2051, 1974.

Kohlstedt, D., C. Goetze and W. Durham, Experimental deformation of single crystal olivine with application to flow in the mantle, in *The Physics and Chemistry of Minerals and Rocks* (ed. S. K. Runcorn) p. 35-49, Wiley, London, 1976.

Liu, H., D. L. Anderson and H. Kanamori, Velocity dispersion due to anelasticity; implications for seismology and mantle composition, *Geophys. J. R. Astron. Soc., 47,* 41-58, 1976.

Minster, J., Transient and impulse responses of a one-dimensional linearly attenuating medium – II. A parametric study. *Geophys. J. R. Astron. Soc., 52,* 508-524, 1978.

Minster, J. B., Anelasticity and attenuation, in Physics of the Earth's Interior (A. Dziewonski and E. Boschi, eds.), *Proc. Enrico Fermi Intern. Sch. Phys.* Academic Press (in press) 1980.

Minster, J. B., and D. L. Anderson, A model of dislocation-controlled rheology for the mantle, in press, *Phil. Trans. Roy. Soc., London,* 1981.

Misener, D. V., Cationic diffusion in olivine to 1400°C and 35 kbar, Carnegie Inst. Washington Publ. 634, 117-129, 1974.

Nowick, A., and B. Berry, Anelastic Relaxation in Crystalline Solids, Academic Press, New York, p. 677, 1972.

Peltier, W., Mantle convection and viscosity, in Physics of the Earth's Interior (A. Dziewonski and E. Boschi, eds.), *Proc. Enrico Fermi Intern. Sch. Phys.,* Academic Press (in press) 1980.

Reddy, K. P. R., S. M. Oh, L. D. Major, Jr., and A. R. Cooper, Oxygen diffusion in forsterite, *J. Geophys. Res., 85,* 322-326, 1980.

Schoeck, G. Friccion interna debido a la interaccion entra dislocaciones y atomos solutos, *Acta Metall. 11,* 617-622, 1963.

Schwenn, M. and C. Goetze, Creep of olivine during hot-pressing, *Tectonophysics, 48,* 41-60, 1978.

Sipkin, S. and T. Jordan, Frequency dependence of Q_{ScS}, *Bull. Seismol. Soc. Amer., 69,* 1055-1079, 1979.

Stocker, R., Influence of oxygen pressure on defect concentrations in olivine with a fixed cationic ratio, *Phys. Earth Planet. Int. 17,* 118-129, 1978.

Stocker, R. and M. Ashby, On the rheology of the upper mantle, *Rev. Geophys. Space Phys., 11,* 391-426, 1973.

Van Bueren, H., Imperfections in Crystals, North-Holland, Amsterdam, p. 676, 1960.

Weertmann, J., Internal friction of metal single crystals, *J. Appl. Phys., 26,* 202-210, 1955.

Weertmann, J., Dislocation climb theory of steady-state creep, *Trans. Am. Soc. Met., 61,* 681-694, 1968.

Woirgard, J., Modèle pour Les pics de frottement interne observés a haute température sur les monocristaux, *Phil. Mag., 33,* 623-637, 1976.

LINEAR VISCOELASTIC BEHAVIOUR IN ROCKS

B.J. Brennan

Department of Physics, University of Auckland, Private Bag, Auckland, New Zealand.

Abstract. The anelastic behavior of rocks and the consequences of this anelasticity are topics of considerable current interest. Problems involving anelasticity are naturally more tractable if the anelastic behaviour is linear, in which case it is described by the theory of linear viscoelasticity. Fortunately there is a growing body of evidence which suggests that at low strain amplitudes this situation applies, at least in the teleseismic frequency range.

For geophysicists one of the most significant results in the theory of linear viscoelasticity is the existence of dispersion relations which relate the frequency dependences of the phase velocity and the attenuation coefficient. The significance of the phase velocity dispersion in the inversion of free oscillation and surface wave data and in comparisions of these data with body wave data has been emphasised in recent years. Cases in which Q is independent of frequency (in the experimental frequency range), or has a power-law dependence on frequency, have been observed in rocks under appropriate physical conditions, and for these simple forms the dispersion may be calculated readily. For both these cases the evolution of a delta function impulse as it propagates is described by a scaling law involving the distance of propagation and the value of the attenuation coefficient at a reference frequency. In particular, the rise time of the pulse has a simple dependence on these quantitites and for the case of constant Q this confirms the rise time relation observed by Gladwin and Stacey.

Recent experiments by the author, in which the stress-strain hysteresis loops for rock specimens subjected to shear were observed directly under laboratory conditions, have demonstrated that under these conditions and at sufficiently small strain amplitudes (typically less than 3×10^{-6}) several rocks exhibit linear anelastic behavior in the frequency range 0.001Hz to 0.5Hz. For two types of igneous rock the Q was very nearly independent of frequency, while for a sandstone specimen Q increased significantly with frequency. In all cases the hysteresis loops were observed to be elliptical at small strain amplitudes, and the variation with frequency of the real part of the shear modulus was in good agreement with that predicted from the observed Q values using the appropriate dispersion relation.

1. Introduction

The term anelasticity describes the dissipation of energy in materials subjected to stress. In the context of seismology this phenomenon is responsible for the attenuation of seismic waves and the damping of free oscillations. It is of fundamental importance to studies involving anelasticity in rocks to establish whether or not the anelastic behavior is linear, in the sense that the principle of superposition of stress signals is valid. It is obvious that non-linear anelastic behavior occurs if the stress and strain amplitudes involved are large enough. In seismology however, the strain amplitudes involved are very small (typically less than 5×10^{-7}) except near the focus of an earthquake. Laboratory experiments have demonstrated that at low temperature and pressure the anelasticity is linear at these strain amplitudes, but the evidence at higher temperature and pressure is not conclusive.

Until recently the evidence for linear behavior at low temperature and pressure also was not conclusive. The most sensitive direct observations of anelasticity in rocks were the experiments of McKavanagh and Stacey (1974), who found that the stress-strain hysteresis loops for rocks subjected to a sinusoidal axial load were cusped at the ends, which is indicative of non-linear anelasticity. However, other results such as those of Pandit and Savage (1973), who observed creep in rocks in flexure at strain levels up to 10^{-5}, were consistent with linear behavior. In addition most observations of pulse or wave propagation in rocks also were consistent with the assumption of linear behavior. Brennan and Stacey (1977), in experiments similar to those of McKavanagh and Stacey (1974), observed that in igneous rock specimens subjected to shear at ambient temperature and pressure the hysteresis loops were elliptical in form, as expected for linear viscoelasticity, and in addition demonstrated that the variation with frequency of the real part of the modulus was of the form predicted by linear theory. The strain amplitudes involved were less than 3×10^{-6}, an order of magnitude smaller than McKavanagh and Stacey (1974) could attain with the same rock types, and the implication is that the earlier experiments involved strain amplitudes at which non-linear mechanisms were operative. This conclusion is supported by the observation by Winkler et al (1979) that in sandstone specimens under small confining pressures (up to 50 bar) the amplitude dependence of the quality parameter Q disappears at low strain amplitudes (typically less than 10^{-6}), which is consistent with linear anelastic behavior at small strain levels. It is significant that at low temperature and pressure non-linear attenuation mechanisms involving open cracks and pores do not appear to contribute to the anelasticity.

Observations of anelasticity in rocks at high temperature and pressure by Goetze (1971), Goetze and Brace (1972) and Murrell and Chakravarty (1973) have all involved large strain amplitudes at which non-linear behavior is evident. Berckhemer et al (1979) recently reported linear anelastic behavior in transient creep experiments on mantle peridotite at a temperature of 1250°C for strain levels up to 5×10^{-5}. However the possibility of partial melting during the experiment casts doubt on the validity of this result as an indicator of anelastic behavior in the mantle. The assumption of linear anelastic behavior is thus much less secure for high sub-solidus temperatures than for low temperatures.

Linear anelastic behavior is described by the theory of linear viscoelasticity. In section 2 several aspects of this theory are summarized. Of particular importance is the phase velocity dispersion which is a consequence of attenuation. The existence of this dispersion requires that measurements made at different frequencies be corrected before comparisions are made. In particular, recognition of the significance of the dispersion has necessitated modications to the procedures for developing earth models from free oscillation and surface wave data (see for example Randall, 1976, or Liu et al., 1976, for details on this topic). The earth model QM2 of Hart et al. (1977) incorporates corrections on the assumption of a frequency-independent Q, and eliminates many of the previous discrepancies between such earth models and body wave data.

The exact nature of the phase velocity dispersion depends on the frequency dependence of the attenuation. In most rock samples at low pressures and room temperature Q is generally very nearly constant in the teleseismic frequency range. However, the arguments of Anderson and Minster (1979), who compared Chandler wobble, tidal and free oscillation data, suggest that at more elevated temperatures and pressures variation of Q as a positive power of frequency may well be appropriate. This conclusion is consistent with the general trend of seismic evidence for frequency dependence of Q, which suggests that Q increases with frequency (see eg Solomon, 1972; Takano, 1971; Toshida and Tsujiura, 1975). Berckhemer et al. (1979) reported variation of Q as a positive power of frequency for mantle peridotite at high temperature, although this result also is subject to doubt due to the possibility of partial melting. A discussion of properties of pulse propagation in media in which Q is constant or varies as a power of frequency in the observed frequency range is included in section 2.

In section 3 results from experiments performed by the author on specimens of sandstone, basalt and granite are presented. These results extend and improve the preliminary work af Brennan and Stacey (1977). The general conclusion of linear behaviour is strengthened. The hysteresis loops are elliptical at small strain amplitudes, and in particular the transition from linear to non-linear behavior is demonstrated for a sandstone specimen . The rock specimens exhibit different variations of Q with frequency, but in general the observed variation with frequency of the real part of the complex shear modulus is in good agreement with the variation predicted from linear theory using the observed Q values.

2. Theoretical Considerations

The results summarized in this section are discussed in more detail by Brennan and Smylie (1981), who review the aspects of the theory of viscoelasticity which are important in seismic applications. The treatment is restricted to consideration of isotropic media only. It should be noted, as indicated by Brennan and Smylie (1981), that under the adiabatic conditions appropriate in seismology the formulation of linear viscoelasticity is indentical in form to the more familiar theory of isothermal linear viscoelasticity, so that the problems of non-linearity encountered in the more general theory of thermoviscoelasticity are fortunately avoided.

In an isotropic linear viscoelastic medium under adiabatic or isothermal conditions the infinitesimal strain and stress tensors are connected by the convolution relations

$$\varepsilon_{kk}(\underset{\sim}{x},t) = \int_{-\infty}^{\infty} C_2(\tau)\, \sigma_{kk}(\underset{\sim}{x},t-\tau)\, d\tau \qquad (1)$$

$$e_{ij}(\underset{\sim}{x},t) = \int_{-\infty}^{\infty} C_1(\tau)\, s_{ij}(\underset{\sim}{x},t-\tau)\, d\tau\ , \qquad (2)$$

where the deviatoric strain and stress tensors are defined by

$$e_{ij} = \varepsilon_{ij} - \frac{1}{3}\,\delta_{ij}\ \varepsilon_{kk} \qquad (3)$$

$$s_{ij} = \sigma_{ij} - \frac{1}{3}\,\delta_{ij}\ \sigma_{kk} \qquad (4)$$

Eqs (1) and (2) represent independent stress-strain relations of form

$$\varepsilon = C*\sigma \ , \tag{5}$$

the star representing convolution.

In Eq. (5) the impulse response or creep rate function C(t) is the (distributional) derivative of the creep function J(t), which is the response to a unit stress imposed at t = 0. In order that the relation between ε and σ be causal it is necessary that J(t) and C(t) be causal functions, that is functions which vanish for $t < 0$. It is generally observed that, for $t > 0$, J(t) is a monotonically increasing function with a monotonically decreasing slope. We call creep functions that satisfy these two conditions and are twice differentiable for $t > 0$ "realistic creep functions". Thus a realistic creep function may be written

$$J(t) = J_o H(t) + \phi(t) \ , \tag{6}$$

where φ(t), the transient creep function, is a condinuous function which is zero for $t < 0$ and satisfies the conditions mentioned above. J_o represents the instantaneous elastic compliance, and may be zero, although in rocks an instantaneous elastic response is indicated. The corresponding creep rate function is

$$C(t) = J_o \delta(t) + \dot{\phi}(t) \ , \tag{7}$$

the dot denoting differentiation with respect to time.

We define the complex Fourier transform $\hat{f}(\xi)$ of a function f(t) by

$$\hat{f}(\xi) = \int_{-\infty}^{\infty} f(t) e^{-i\xi t} \, dt \ , \tag{8}$$

with

$$\xi = \omega + i\lambda \ . \tag{9}$$

The transforms $\hat{C}(\xi)$ and $\hat{J}(\xi)$ are causal transforms, that is they are analytic in the lower half plane (l.h.p.) $\lambda < 0$, as a consequence of the causal nature of C(t) and J(t), and because of the reality of these functions their transforms have the Hermitian property

$$\hat{C}(-\bar{\xi}) = \bar{\hat{C}}(\xi) \ , \tag{10}$$

the bar denoting complex conjugation.

The Fourier transform of Eq. (5) is

$$\hat{\varepsilon}(\omega) = \hat{C}(\omega) . \, \hat{\sigma}(\omega) \ , \tag{11}$$

and is valid with argument ξ in the l.h.p., in particular for causal stress signals. The complex compliance $\hat{C}(\omega)$ is the analogue of the normal compliance in elasticity theory, and is the reciprocal of the modulus $\hat{M}(\omega)$. The requirement that the net work done in deforming a viscoelastic material from the unstressed state must be non-negative, so that the material is passive, leads to the result that the transform $\hat{C}(\xi)$, or alternatively $\hat{G}(\xi)$, the transform of the relaxation function G(t), is a positive real function, that is a causal transform, the real part of which is non-negative throughout the l.h.p. Conversely, all positive real functions correspond to passive systems. One consequence of this is that the transforms of J(t) and G(t) and their derivatives have no zeros in the l.h.p. and are thus minimum phase transforms. Another consequence is that the imaginary part of $\hat{C}(\omega)$ has the opposite sign to ω. All realistic creep functions can be shown to yield the positive real property, and thus form a subset of the set of creep functions that are permissible.

By considering the equations of motion in a homogeneous, isotropic, viscoelastic medium, one obtains equations which are equivalent to the Navier equations of standard elasticity theory. In the frequency domain they are identical in form, with the constant real moduli of elasticity theory replaced by complex moduli which depend on frequency. Plane wave solutions to those equations may be obtained for shear and compressional waves with complex wave numbers

$$\kappa_S(\omega) = \omega\sqrt{\rho/\hat{\mu}(\omega)} \tag{12}$$

and

$$\kappa_P(\omega) = \omega\sqrt{\rho/(\hat{K}(\omega) + 4\hat{\mu}(\omega)/3)} \tag{13}$$

respectively. The shear modulus $\hat{\mu}$ and bulk modulus $\hat{K}$ are defined by

$$\hat{\mu} = (2\hat{C}_1)^{-1} \tag{14}$$

$$\hat{K} = (3\hat{C}_2)^{-1} \ . \tag{15}$$

It may be shown that the analytic continuation of both wave numbers is analytic and minimum phase in the l.h.p. As a consequence dispersion relations of the Kramers-Kronig type relate the real and imaginary parts of each wave number on the real axis. Since

$$\kappa_{S,P}(\omega) = \omega/v(\omega) - i\beta(\omega) \ , \tag{16}$$

the dispersion relations relate the phase velocity v and the attenuation coefficient β. If c is the high frequency limit of v (which may classically be infinite if J_o in Eq. (6) is zero), then the dispersion relations may be obtained by considering the integral of the analytic function

$$h(\xi) = (\kappa(\xi)/\xi - 1/c)/(\xi - \omega_o) \tag{17}$$

around a semicircular contour in the l.h.p. Using the fact that both v and β are even functions of ω, these Kramers-Kronig relations may be written

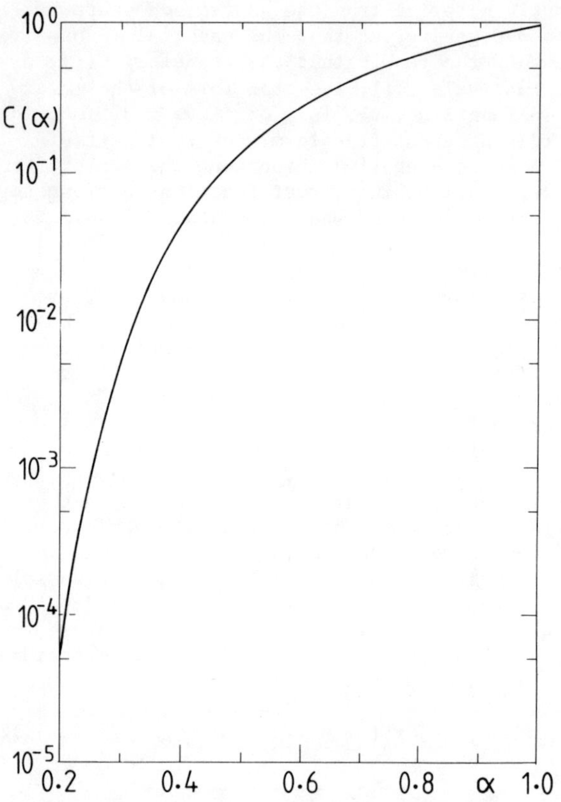

Fig.1. Graph showing the variation of the rise time parameter $C(\alpha)$ in the relation $\tau = C(\alpha)\ (\int B dx)^{1/\alpha}$ with the parameter α in the relation $\beta = B|\omega|^{\alpha}$.

$$\frac{1}{v(\omega_o)} = \frac{1}{c} - \frac{2}{\pi}\, pv \int_o^{\infty} \frac{\beta(\omega)}{\omega_o^{\,2}-\omega^2}\, d\omega \qquad (18)$$

$$\beta(\omega_o) = \frac{2\omega_o^{\,2}}{\pi}\, pv \int_o^{\infty} \frac{1}{v(\omega)\,(\omega_o^{\,2}-\omega^2)}\, d\omega \qquad (19)$$

in which the principal values of the integrals are taken. The existence of the dispersion relations requires that $\beta(\omega)/\omega$ tend to zero as ω tends to ∞. The class of realistic creep functions satisfies this condition. In considering pulse propagation it may be shown that the onset of a pulse propagates with speed c. However, for media with realistic creep functions $v(\omega)$ is always smaller then c and thus, depending on the spectral content of the pulse, the onset of a measurable signal may be somewhat later than the theoretical arrival time.

The dispersion relation of Eq. (18) has in practice been the more useful one in seismological applications. As mentioned in the introduction, the Q of rocks at low pressure and ambient temperature tends to be very nearly independent of frequency, while higher temperature experiments suggest that Q may vary as a positive power of frequency, and this latter form may well be appropriate in much of the interior of the earth. It is thus appropriate to examine the nature of the dispersion associated with such forms for Q and the properties of pulse propagation in media in which these forms apply.

We define Q by the relation

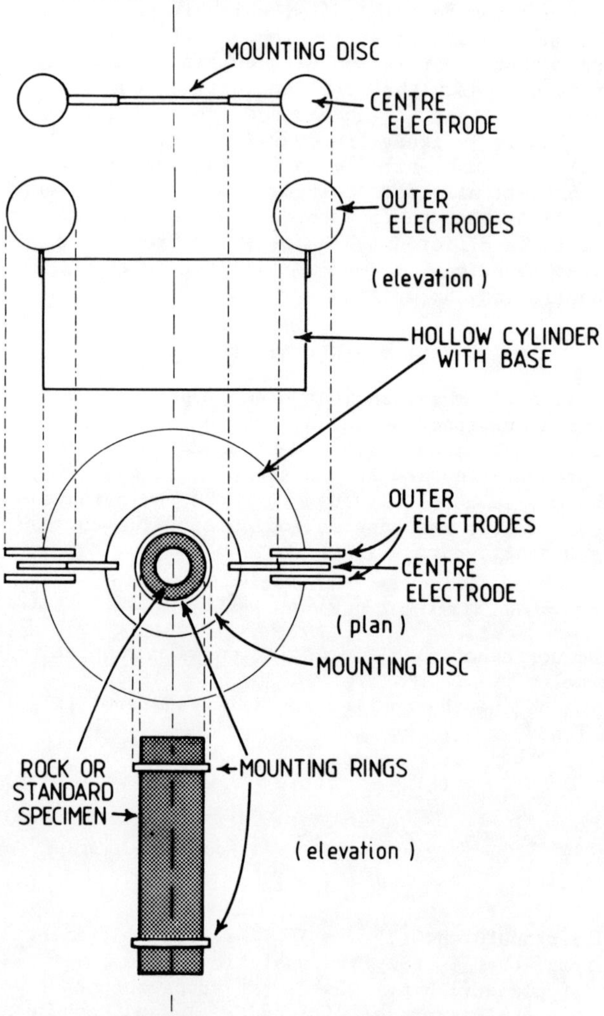

Fig.2. Schematic diagram of a specimen and the associated electrode system for the measurement of strain. The centre electrodes are attached to a mounting disc which is screwed onto the upper mounting ring on the specimen. The outer electrodes are attached to backing plates (not shown) mounted on a hollow cylinder, the base of which is screwed onto the lower mounting ring on the specimen. Under torsion the movement of the centre electrode of each of the three electrode systems with respect to the outer electrodes is proportional to the shear strain in the specimen, and may be measured using a ratio transformer bridge.

$$Q(\omega) = |\omega|/2\beta(\omega)v(\omega) \ . \tag{20}$$

Since the phase velocity dispersion is small for Q of order 100 or more (certainly no more than 5% over 3 decades in frequency), Q proportional to $|\omega|^{1-\alpha}$ implies that β is very nearly proportional to $|\omega|^{\alpha}$. The case of constant Q may be considered as the limit as α tends to unity. If

$$\beta(\omega) = B|\omega|^{\alpha}, \quad 0 < \alpha < 1 \ , \tag{21}$$

so that to a high degree of approximation

$$Q = Q_1 \ |\omega|^{1-\alpha} \ ,$$

then by Eq. (18)

$$v(\omega)^{-1} = c^{-1} + B\tan(\alpha\pi/2) \ |\omega|^{\alpha-1} \ , \tag{22}$$

which in the constant Q limit ($\alpha \to 1$) becomes

$$v(\omega)^{-1} = v(\omega_r)^{-1} - 2B\ln|\omega/\omega_r|/\pi \tag{23}$$

with ω_r a reference angular frequency. If Q_1 is 100, then the dispersion over the frequency range 0.002Hz to 2Hz is 2.2% by Eq. (23), and by Eq.(22) 3.2% and 4.3% when α is 0.7 and 0.5 respectively.

Brennan (1980) has shown that, in a homogeneous viscoelastic medium in which Eqs (21) and (22) apply, the propagation in the positive x direction of a plane waveform which is initially a delta function stisfies the simple scaling law

$$u(x,t) = (Bx)^{-1/\alpha} \ U_{\alpha}(\theta) \ , \tag{24}$$

where

$$t = x/c + (Bx)^{-1/\alpha} \ \theta \tag{25}$$

and

$$U_{\alpha}(\theta) + \pi^{-1} \int_0^{\infty} \exp(-\omega^{\alpha}) \ \cos[\omega\theta - \omega^{\alpha}\tan(\alpha\pi/2)]d\omega. \tag{26}$$

If B and c, but not α, vary with x, then x/c and Bx are replaced by $\int dx/c$ and $\int Bdx$ respectively.

The amplitude of the propagating pulse in Eq. (24) varies as $(\int Bdx)^{-1/\alpha}$ and the pulse rise time τ, defined as in Gladwin and Stacey (1974) by

$$\tau = u_{max}/(\partial u/\partial t)_{max} \ , \tag{27}$$

varies as

$$\tau = C(\alpha) \ (\int Bdx)^{1/\alpha} \tag{28a}$$

$$\cong \tfrac{1}{2}C(\alpha) \ (\int dt/Q_1)^{1/\alpha} \ , \tag{28b}$$

with

$$C(\alpha) = (U_{\alpha})_{max}/(\partial U_{\alpha}/\partial\theta)_{max} \ . \tag{29}$$

The variation of $C(\alpha)$ with α is shown in Fig.1. For α between 0.2 and 1, $C(\alpha)$ may be approximated to within 4% by the expression

$$C(\alpha) = \exp(1.661 - 1.500\alpha^{-1} - 0.164\alpha^{-2}). \tag{30}$$

In the constant Q limit

$$\tau = 0.97 \ (\int Bdx) \ . \tag{31}$$

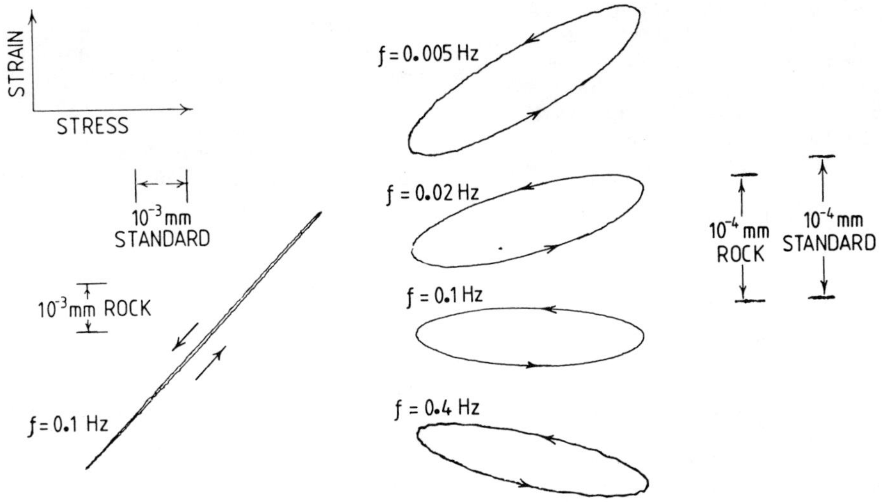

Fig.3. Hysteresis loops for a sandstone specimen at a strain amplitude of 1.2×10^{-6}. The direct plot of strain against stress is shown at the left, and expanded loops at different frequencies are shown at the right. The calibration marks are obtained by switching the appropriate decade of the ratio transformer bridge. The elliptical form of the expanded loops is apparent, and dispersion is demonstrated by the change in slope with frequency for these loops.

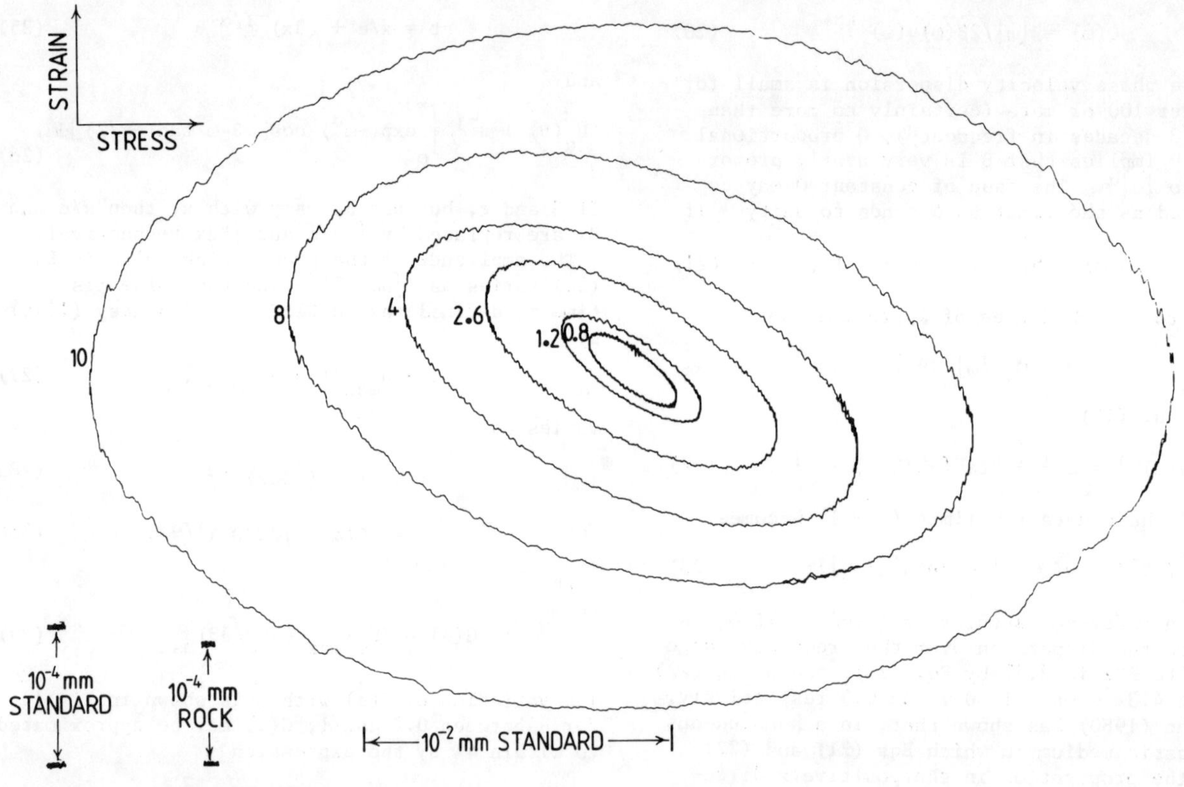

Fig.4. Expanded hysteresis loops for a sandstone specimen at a frequency of 0.1Hz with strain amplitudes varying between 8×10^{-7} and 10^{-5}. The figure by each loop denotes the maximum microstrain for that loop. For strain amplitudes below 1.2×10^{-6} the loops are indistinguishable from ellipses, but at higher amplitudes non-linear effects contribute.

In the constant Q case the experimental results of Gladwin and Stacey (1974) indicate that an equation of the same form as Eq. (31) describes the change with propagation of τ for the first arrival of a wavetrain when the initial rise time τ_o at $x = 0$ is relatively small. Their constant of 1.06 ± 0.08 agrees quite well with the constant 0.97 in Eq. (31). The appropriate generalisation of Eq. (28a) is

$$\tau^{\alpha} = \tau_o^{\alpha} + C^{\alpha}(\alpha) \; (\int B dx) \; . \qquad (32)$$

It is conceivable that Eq. (32) may be of value in determining variations of average B or Q_1 with path and thereby elucidating the Q structure of the earth. This rise time technique requires a shorter record and is not as strongly affected by multiple arrivals as are spectral ratio techniques for determining Q.

In using Eq. (32) it is necessary to assume a value of α. The arguments of Anderson and Minster (1979) suggest that a value of about 0.7 is appropriate, and this value agrees closely with the value of 0.73 reported by Berckhemer et al. (1979) for peridotite. Jeffreys and Crampin (1970) and Jeffreys (1976) obtained higher estimates of 0.81 and 0.87 respectively by comparing body wave and Chandler wobble data.

Experimental Results

The results of experiments performed by the author while at the Department of Physics at the University of Queensland are presented in this section. The experiments observed directly the stress-strain hysteresis loops in sandstone, basalt and granite rock specimens subjected to sinusoidal torsion, and were similar in principle to those of McKavanagh and Stacey (1974) The apparatus used was capable of observing fine details of loop shape at a strain amplitude of 10^{-6} or less, considerably lower than could be achieved in previous direct observations of anelasticity in rock samples.

The rock specimens used were tubes of length approximately 280 mm, external diameter 44 mm and wall thickness 5 mm. Tubes rather than solid cylinders were used to ensure that the strain, which was pure shear, was nearly uniform throughout the specimens. A standard specimen of either high Q steel or fused silica was subjected to the same torque as the rock specimen, and the strain on the standard specimen was used to infer the stress in the rock specimen. Tests in which the two standard specimens were compared indicated that the Q of one relative to the other (ie $(Q_1^{-1} - Q_2^{-1})^{-1}$) was at least 4000 for frequencies between 0.002Hz and 0.2Hz. It appeared that the steel standard has the higher Q, presumably of order 5000 or more, and this standard was used to obtain the results presented here.

Two three-electrode capacitance displacement transducers (Stacey et al. 1969; Gladwin and Wolfe, 1975) were used to measure the strain in each specimen. Figure 2 shows schematically the arrangement used. Threaded mounting rings were attached about 20 mm from both ends of the specimen. The centre electrodes of the two three-electrode sets were mounted diametrically opposite on a mounting disc screwed onto the upper mounting ring, and the outer electrodes were mounted on backing plates (not shown in Fig. 2) which were attached to a hollow cylinder, the base of which was screwed onto the lower mounting ring. In each electrode set the outer electrodes were rigidly fixed together and adjusted so that at equilibrium the gap between the centre electrode and each outer electrode was very nearly 0.5 mm. The electrode sets were centred about 200 mm from the axis of the specimen. If the specimen is subjected to torsion the centre electrodes and their mounting disc move with

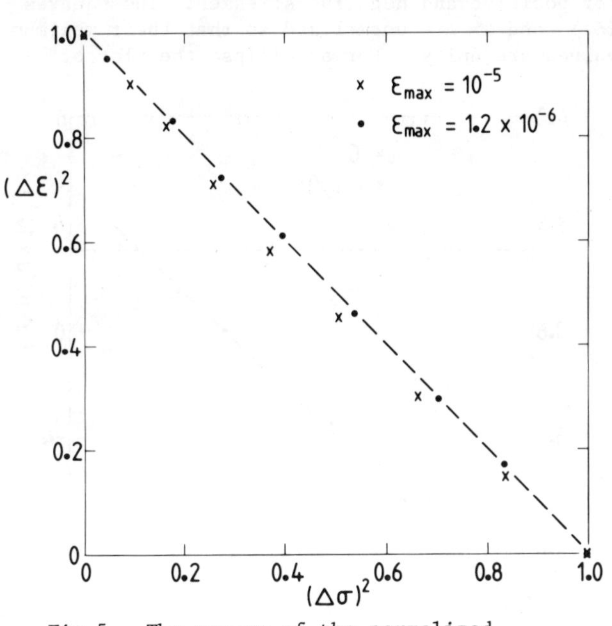

Fig.5. The square of the normalized strain difference between the upper and lower arms of the expanded hysteresis loop plotted against the square of the normalized stress for a sandstone specimen at two different strain amplitudes. An elliptical loop satisfies the linear relationship represented by the broken line. The fit to this line is excellent at a strain amplitude of 1.2 x 10^{-6}, but non-linear effects are evident at a strain amplitude of 10^{-5}.

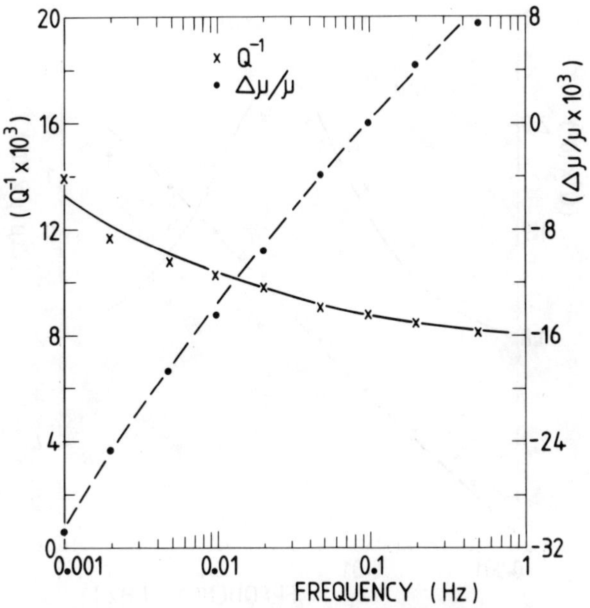

Fig.6. Plot of Q^{-1} and $\Delta\mu/\mu$ against frequency for a sandstone specimen. The Q^{-1} and $\Delta\mu/\mu$ values have typical standard errors of 0.16 x 10^{-3} and 0.4 x 10^{-3} respectively. The solid line represents the best fitting curve $Q^{-1} = A + Bf^C$ and the broken line shows the expected variation of $\Delta\mu/\mu$ when Q^{-1} has this form.

respect to the outer electrodes attached to the cylinder, the relative displacement being proportional to the strain in the specimen. The two capacitances in each electrode set may be connected to a ratio transformer bridge with a rapid response time (10^{-3} s). At balance the bridge gives a seven figure reading, which is the ratio of one capacitance gap to the sum of the two gaps (1 mm) and off balance the detector circuit receives a signal proportional to the displacement from the balance position (see Stacey et al., 1969, for details). In practice diagonally opposite capacitances were connected in parallel, thereby reducing the effects of asymmetry, and in particular suppressing any signal due to flexural rather than torsional strains. The effects of thermal expansion were minimised by mounting the backing plates for the outer electrodes on the same side in the plan view.

The rock and standard specimens were mounted one on top of the other and fixed at the base. A small torque was applied to the top of the upper specimen through two loudspeaker coil movements mounted diametrically opposite one another and attached to the specimen by lever arms of length approximately 180 mm. The coils were driven by a low frequency sinusoidal oscillator and the out of balance signals from the bridges for the two specimens were recorded on an X-Y recorder, giving in effect a plot of strain against stress. For a

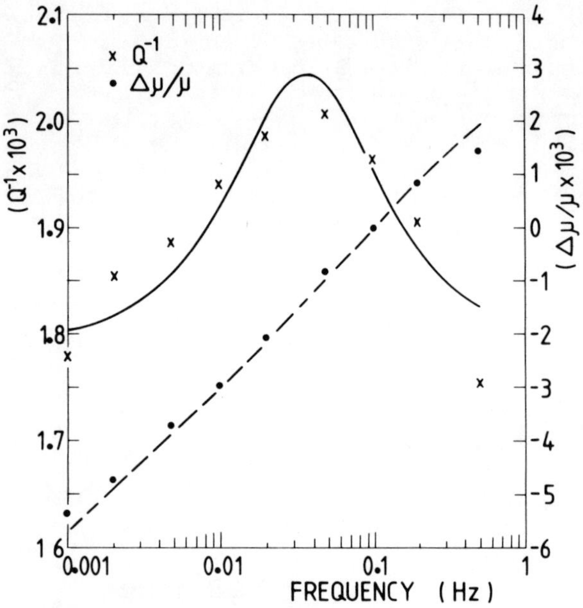

Fig.7. Plot of Q^{-1} and $\Delta\mu/\mu$ against frequency for a basalt specimen. The Q^{-1} and $\Delta\mu/\mu$ values have typical standard errors of 0.023×10^{-3} and 0.07×10^{-3} respectievely. The solid line represents the best fitting curve $Q^{-1} = A + Bf/(1 + Cf^2)$ and the broken line shows the expected variation of $\Delta\mu/\mu$ when Q^{-1} has this form.

specimen with high Q, the hysteresis loop is barely apparent but by subtracting the appropriate proportion of the standard signal from the rock signal and further amplifying the difference, one may plot directly the departure of rock strain from linear elasticity against the strain of the standard. The area of the expanded loop is a measure of the hysteretic loss in the rock specimen relative to the standard specimen. Q values may be calculated using the relation

$$Q^{-1} = \Delta E/2\pi E , \quad (33)$$

where E is the peak stored elastic energy per unit volume (approximately $\frac{1}{2}\varepsilon_{max}\cdot\sigma_{max}$) and ΔE is the energy loss per cycle per unit volume (ie the area of the loop). The slop of the major axis of the expanded loop is a sensitive indicator of changes in the real part of the viscoelastic modulus.

Figure 3 shows direct and expanded loops obtained using a sandstone specimen at a strain amplitude of 1.2×10^{-6}. Note that for the expanded loops the gain on the strain axis is about 30 times that for the direct loop. The expanded loops demonstrate the elliptical shape of the hysteresis loops at this strain level, and the variation with frequency of the real part of the modulus, and therefore of the phase velocity, is demonstrated by the change in slope of the major axis of the ellipse. Note that for small changes in modulus the changes in velocity and modulus are related by

$$\frac{v(\omega)-v(\omega_r)}{v(\omega_r)} = \frac{1}{2}\,\frac{\mu(\omega)-\mu(\omega_r)}{\mu(\omega_r)} , \quad (34)$$

with ω_r being a reference angular frequency.

The transition from linear to non-linear behaviour in the sandstone specimen, as the strain amplitude increases, is shown in Figs 4 and 5. Fig.4 shows expanded loops obtained for different strain amplitudes between 8×10^{-7} and 10^{-5} at a frequency of 0.1Hz. Q^{-1}, as measured using Eq. (33), increases with the strain amplitude and is almost 30% larger at an amplitude of 10^{-5} than for strain amplitudes of order 1.2×10^{-6} or less. This amplitude dependence of Q^{-1} is indicative of non-linear behaviour at the larger strain amplitudes. The loop shape at the higher strain amplitudes also deviates from the elliptical shape expected for a linear material, as demonstrated by Fig. 5. $\Delta\varepsilon$ is the difference in strain between the upper and lower arms of the hysteresis loop for a given stress σ, averaged for positive and negative stresses. The squares $(\Delta\varepsilon)^2$ and σ^2 are normalised so that their maximum values are unity. For an ellipse the plot of

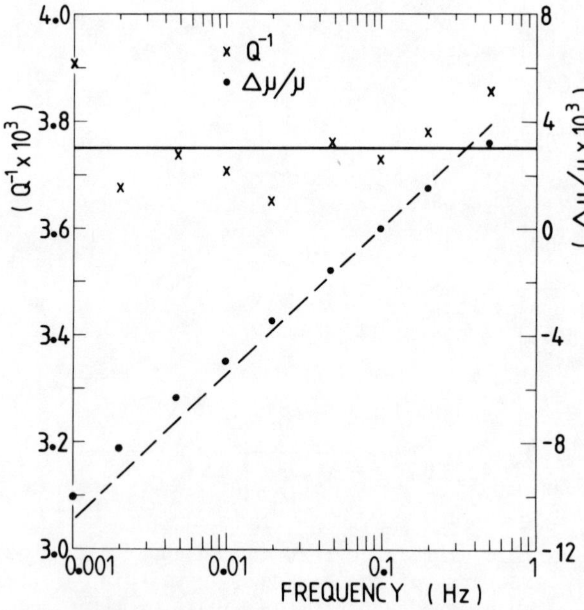

Fig.8. Plot of Q^{-1} and $\Delta\mu/\mu$ against frequency for a granite specimen. The Q^{-1} and $\Delta\mu/\mu$ values have typical standard errors of 0.07×10^{-3} and 0.13×10^{-3} respectively. The solid line represents $Q^{-1} = 0.00375$, and the broken line shows the expected variation of $\Delta\mu/\mu$ for this value of Q^{-1}.

$(\Delta\varepsilon)^2$ against σ^2 is a straight line, and the data for a strain amplitude of 1.2×10^{-6} fit this line extremely well. At a strain amplitude of 10^{-5} it can be seen that the loop is more pointed at the ends than an ellipse. This is consistent with the observations of McKavanagh and Stacey (1974), though their hysteresis loops differed from ellipses rather more dramatically than the loops presented here. Similar tests of loop shape for the basalt and granite specimens indicate that these specimens exhibit linear behavior at strain amplitudes up to 3×10^{-6}.

Hysteresis loops were observed in all three types of rock at frequencies between 0.001Hz and 0.5Hz. Figures 6, 7 and 8 show the averages of the values of Q^{-1} and $\Delta\mu/\mu$ (ie the right hand side of Eq. 34) measured from these loops. The reference frequency used was 0.1Hz.

The results for a sandstone specimen at a strain level of 1.2×10^{-6} are shown in Fig. 6. This specimen exhibited a quite strong frequency dependence of Q^{-1}, and in addition Q^{-1} was observed to vary with weather conditions, in particular to increase with humidity. The data points in Fig. 6 represent the average of values obtained in three complete runs for which Q^{-1} had similar values. The solid line is the best fitting curve obtained by assuming Q^{-1} to be a sum of constant and power-law terms, viz;

$$Q^{-1} = 0.00702 + 0.00934\ f^{-0.275}\ . \qquad (35)$$

The dotted line represents the variation of $\Delta\mu/\mu$ to be expected when Q^{-1} has this form, as calculated using Eqs (18), (20) and (34). The agreement with the measured values of $\Delta\mu/\mu$ is excellent, and provides further confirmation of the linear behaviour of the sandstone specimen at this strain level.

Fig. 7 shows the results for a basalt specimen at a strain amplitude of 1.7×10^{-6}. This specimen has a high Q (about 520), and exhibits a small peak in the Q^{-1} curve in the observed frequency range. The solid curve, which represents a moderate fit to the Q^{-1} values, is the curve

$$Q^{-1} = 0.00179 + 0.0000183f/(0.00129 + f^2) \qquad (36)$$

obtained by assuming that Q^{-1} is the sum of a constant and a "standard linear solid" resonance peak. The dotted line again represents the variation of $\Delta\mu/\mu$ predicted by the dispersion relation. The agreement with the observed values is quite good, though the average slope of the predicted curve is slightly greater than that for the observed values. The best fitting straight line through the observed values suggests a constant Q^{-1} of 0.00183.

The results for a granite speciment at a strain amplitude of 1.1×10^{-6} are shown in Fig. 8. This specimen was more prone to drift problems than the other two so that the data are less precise. The Q is reasonably high (about 270) and appears fairly constant over the observed frequency range, with an average value for Q^{-1} of 0.00375. The variation of $\Delta\mu/\mu$ predicted from this value is shown as the dotted straght line. While the experimental values do fit a straight line quite well, as expected in a constant Q material, the gradient does not agree precisely with the predicted gradient, and suggests a value for Q^{-1} of 0.00335. Linear anelastic effects in the apparatus or the standard specimen do not affect the expected relationship between the observed values of Q^{-1} and $\Delta\ell n\mu/\mu$, and cannot explain the discrepancy. The loops observed at the reference frequency of 0.1Hz are elliptical in form, indicating linear behavior. It is possible that a non-linear mechanism, possibly associated with the drift observed, and not necessarily in the rock specimen itself, contributes at lower frequencies. However, tests on loops at low frequencies have indicated that they are elliptical within the precision of the measurements.

4. Conclusion

It has been usual in seismic studies involving anelasticity in the Earth to assume linear behavior. The assumption of linearity is naturally of considerable convenience, as even simple problems in anelasticity become virtually intractable if non-linear theories must be used. If linearity applies, then the results of the theory of linear viscoelasticity such as those presented in section 2 apply, and many problems are identical in form to the corresponding elastic problem with frequency-dependent complex moduli replacing the constant real moduli of elasticity. With the exception of the discrepancy in the slope of the $\Delta\mu/\mu$ curve for granite, the results presented in section 3 have provided strong evidence that the linearity assumption is justified for most rock types at low temperature and pressure. The extrapolation of this assumption to the higher pressures and temperatures encountered in the interior of the Earth remains tentative, in the absence of reliable data on rock anelasticity at small strain amplitudes and elevated temperature and pressure.

Acknowledgements. The experimental work was supported by a grant from the Australian Research Grants Committee to Professor F.D. Stacey, who also provided valuable assistance and comments. Mr R. Willoby constructed the mechanical apparatus and Dr D. Hainsworth and Mr G. Dunne designed and built the electronic circuitry for the ratio transformer bridges. Professor D.E. Smylie made valuable contributions to my theoretical understanding of anelasticity.

References

Anderson, D.L., and J.B. Minster, The frequency dependence of Q in the Earth and implications for mantle rheology and Chandler wobble,

Geophys. J. R. Astr. Soc., 58, 431-440, 1979.
Berckhemer, H., F. Auer, and J. Drisler, High temperature anelasticity and elasticity of mantle peridotite, Phys. Earth Planet. Int., 20, 48-59, 1979.
Brennan, B.J., Pulse propagation in media with frequency-dependent Q, Geophys. Res. Lett., 7, 211-213, 1980.
Brennan, B.J., and D.E. Smylie, Linear viscoelasticity and dispersion in seismic wave propagation, Rev. Geophys. Space Phys., in press.
Brennan, B.J., and F.D. Stacey, Frequency dependence of elasticity of rock - test of seismic velocity dispersion, Nature, 268, 220-222, 1977.
Gladwin, M.T., and F.D. Stacey, Anelastic degradation of acoustic pulses in rock, Phys. Earth Planet. Int., 8, 332-336, 1974.
Gladwin, M.T., and J. Wolf, Linearity of capacitance displacement transducers, Rev. Sci. Instrum., 46, 1099-1100.
Goetze, C., High temperature rheology of Westerly granite, J. Geophys. Res., 76, 1223-1230, 1971.
Goetze, C., and W.F. Brace, Laboratory observations of high-temperature rheology or rocks, Tectonophysics, 13, 583-600, 1972.
Hart, R.S., D.L. Anderson, and H. Kanamori, The effect of attenuation on gross earth models, J. Geophys. Res., 82, 1647-1654, 1977.
Jeffreys, H., The damping of P waves, Geophys. J. R. Astr. Soc., 47, 347-349, 1976.
Jeffreys, H., and S. Crampin, On the modified Lomnitz law of damping, Mon. Not. R. Astr. Soc., 147, 295-301, 1970.
Liu, H.P., D.L. Anderson, and H. Kanamori, Velocity dispersion due to anelasticity; implications for seismology and mantle composition, Geophys. J. Roy. Astron., 47, 41-58, 1976.
McKavanagh, B., and F.D. Stacey, Mechanical hysteresis in rocks at low strain amplitudes and seismic frequencies, Phys. Earth Planet Int. 8, 246-250, 1973.
Murrell, S.A.F., and S. Chakravarty, Some new rheological experiments on igneous rocks at temperatures up to 1120°C, Geophys. J. Roy. Astron. Soc., 34, 211-250, 1973.
Pandit, B.I., and J.C. Savage, An experimental test of Lomnitz's theory of internal friction in rocks, J. Geophys. Res., 78, 6097-6099, 1973.
Randall, M.J., Attenuative dispersion and frequency shifts of the Earth's free oscillations, Phys. Earth Planet. Int., 12, P1-P4, 1976.
Solomon, S.C., On Q and seismic discrimination, Geophys. J. Roy. Astron. Soc., 31, 163-177, 1972.
Stacey, F.D., J.M.W. Rynn, E.C. Little, and C. Croskell, Displacement and tilt transducers of 140dB range, J. Phys. E. (J. Sci. Instrum.), 2, 945-949, 1969.
Takano, K., A note on the attenuation of short period P and S waves in the mantle, J. Phys. Earth., 19, 155-163, 1971.
Winkler, K., A. Nur, and M. Gladwin, Friction and seismic attenuation in rocks, Nature, 277, 528-531, 1979.
Yoshida, M., and M. Tsujiura, Spectrum and attenuation of multiply reflected core phases, J. Phys. Earth., 23, 31-42, 1975.

DIFFERENTIAL ATTENUATION COEFFICIENTS

FOR RAYLEIGH WAVES: A NEW CONSTRAINT ON Q-MODELS

Michel Cara

Seismological Laboratory, California Institute of Technology

Pasadena, California 91125

Abstract. The relative variations of the attenuation coefficients $\delta\gamma_n(T)$ with respect to the period T and the rank of the mode n can be estimated from observed amplitudes provided that the relative excitation of the modes is known at the epicenter. Such a method is applied to the fundamental and the first two higher Rayleigh modes observed in the continental United States for 4 intermediate-depth earthquakes located in the New Hebrides. In the period range 30-100s, the measurement of higher Rayleigh mode amplitudes requires use of both multiple frequency filtering and spatial filtering of records observed on a wide array of long period stations in order to avoid large errors due to interferences between modes. In a first approximation, the coefficients $\delta\gamma_n(T)$ depend linearly on the anelastic parameters within the Earth. The observed values, mainly related to the Pacific Ocean are compatible with a frequency-independent but depth-dependent Q model of the upper 700 km of the Earth. If all the elastic energy loss occurs in shear, the data require an average Q_s value of about 80-100 between 100 and 300 km and a higher average Q_s above 100 km and between 300 and 700 km. Two similar tentative models are proposed for the upper 700 km of the Earth which exhibit an average Q_s value greater than 180 between 250 and 700 km.

Introduction

The anelastic properties of the upper mantle have received considerable attention during these last years. A better understanding of the attenuation behavior of seismic waves is important not only for a better knowledge of the Earth material but also because the physical dispersion associated with the attenuation should be taken into account in many seismological studies, in particular when the low Q zone of the upper mantle is involved (Liu et al., 1976; Kanamori and Anderson, 1977).

The presently available seismological data poorly constrain the Q values of both P and S waves in the mantle. Indeed, very simple frequency independent Q models fit the normal mode attenuation observations (e.g., Sailor and Dziewonski, 1978). Anelastic models with spherical symmetry are still widely used for Q investigations in the mantle, except for the uppermost 100-150 km where constraints may be obtained from regional surface wave data (Solomon, 1972; Mitchell et al., 1976, 1977).

Among many problems, one important question to be addressed is that of the frequency-dependence of Q in the low Q regions of the upper mantle. Although frequency-independent Q models are generally sufficient to explain the seismological data, seismological evidence for a slight frequency dependence of Q have been recently reviewed by Anderson and Minster (1979).

Single-mode surface wave data are not appropriate to investigate the frequency dependence of Q within the mantle since the depth-range over which the Q information is taken depends on frequency. On the other hand knowledge of the attenuation properties of higher modes in a period range 30-100s would be of great importance to investigate both the frequency and the depth-dependence of Q, since for a fixed frequency different modes sample the upper mantle over a depth-range which depends on the rank of the mode.

To date, only few estimates of the attenuation of higher modes have been reported in the literature. In the period-range 30-100s the only estimates to our knowledge are from Okal (1979) for the first and second higher Rayleigh modes. Other estimates of higher mode Q have been obtained at longer periods from free oscillation data (Jobert and Roult, 1976; Sailor and Dziewonski, 1978).

In the period range 30-100s the first and second higher Rayleigh modes may be isolated from a set of long period records observed on a wide array of stations by using pass-band phase velocity filters - or spatial filtering (Cara, 1978). Previous attempts to obtain attenuation data for higher modes within an array of stations have provided data exhibiting considerable scatter (Cara, 1978).

As an alternative to study the attenuation of Rayleigh modes, one can estimate the variation of the spatial attenuation coefficients not within an array of stations but for the full path between an epicenter and an array of stations. By doing that, the length of the path over which the surface waves are attenuated may be significantly increased and a better estimation of the attenuation may be expected from the observed amplitudes. On the other hand, the method requires accurate knowledge of the excitation of the waves at the source and the uncertainty in the source parameters is a new cause of errors.

A similar technique had been applied by Tsai and Aki (1970) to infer the focal depth from Rayleigh waves. For near surface earthquakes and periods lying between 15 and 50s, these authors have shown that the most sensitive focal parameter is the depth of the earthquake. Provided that the attenuation coefficient is assumed to be known, the focal depth may then be estimated from the shape of the observed Rayleigh or Love wave spectrum. In fact uncertainty in the attenuation coefficient, mainly when sedimentary oceanic basins are involved, severely limited the application of this method (Weidner, 1975).

The events we used in this study are intermediate depth earthquakes of magnitude 6 or greater. For these events, both the focal depth and the focal mechanism may be well constrained from observed teleseismic body waves. If sufficient accuracy is reached in the estimation of the focal parameters one can in turn infer the variation of the attenuation coefficient versus period and the rank of the mode as shown below. The measurement of the absolute value of the attenuation coefficients would necessitate an independent estimate of the seismic moment of the source.

Measurement of $\delta\gamma_{no}(T)$ from Observed Amplitudes

The theoretical amplitude $A_n^t(T)$ at a distance x from the source for a mode n and a period T may be written as:

$$A_n^t(T) = I^t(T)\, \mathcal{M}^t\, F_n^t(T) \exp\left(-\gamma_n^t(T)x\right)/\sqrt{x} \quad (1)$$

where $I^t(T)$ is computed instrumental response, $\mathcal{M}^t$ a hypothetical seismic moment and $\gamma_n^t(T)$ a hypothetical attenuation coefficient for mode n at period T.

For a point double couple source, the source amplitude spectrum $F_n^t(T)$ of the vertical motion is (Ben Menahem and Harkrider, 1964):

$$F_n^t(T) = S^t(T)\, \chi_n^t(\Theta^t)\, A_n^t(T) \sqrt{k_n^t} \quad (2)$$

where $S^t(T)$ is the Fourier transform of the time source function, A_n^t is the amplitude response for the Earth model used in the computations, k^t is the wave number and $\chi_n^t(\Theta^t)$ is a function depending on the azimuth Θ^t of the station with respect to the strike direction, on the orientation of the double-couple and on the values of the stress and longitudinal displacement at the depth of the source.

Neglecting the noise in the data, the actual spectrum observed at the epicentral distance x may be written as

$$A_n(T) = I(T)\, \mathcal{M}\, F_n(T) \exp\left(-\gamma_n(T)\, x\right), \quad (3)$$

where

$$F_n(T) = S(T)\, \chi_n(\Theta)\, A_n(T) \sqrt{k_n}.$$

In the above expressions, $\mathcal{M}$ is the actual seismic moment and all the quantities written with an index t in (1) and (2) (theoretical values) are written without index in (3) (actual values). Of course by assuming that $A_n(T)$ in (3) is the observed amplitude of the mode n we suppose that: (1) the noise level for a single-mode amplitude measurement is negligible; (2) the point double couple model of the source is valid. The first assumption may be approached by applying both spatial filtering and multiple frequency filtering to the records (Cara, 1978). However, we shall see that residual interference effects between modes remain a source of error, even after application of both kind of filters. The second assumption is reasonable for the events used in this study as we shall see in the "Data Analysis" section. b. Smaller earthquakes would be difficult to use due to uncertainty in the focal parameters and stronger earthquakes would necessitate a more detailed study of the source due to the large spatial event of such sources.

One can estimate from the data the differential quantity:

$$\varepsilon_{no}(T) = \mathrm{Ln}\left(A_n(T)/A_n^t(T)\right) - \mathrm{Ln}\left(A_o(T_o)/A_o^t(T_o)\right)$$

where o designs a reference mode (the fundamental mode hereafter) and a reference period. $\varepsilon_{no}(T)$ is related to the relative attenuation coefficient

$$\delta\gamma_{no}(T) = \gamma_n(T) - \gamma_o(T_o) \quad (4)$$

by the relation

$$\delta\gamma_{no}(T) = -\varepsilon_{no}(T) + \delta\eta_{no}(T) + \gamma_n^t(T) - \gamma_o^t(T_o) \quad (5)$$

where the "error" term $\delta\eta_{no}(T)$ is

$$\delta\eta_{no}(T) = \mathrm{Ln}\,\Big(I(T)\, S(T)\, \chi_n(\Theta)\, A_n(T) \sqrt{k_n} / I^t(T)\, S^t(T) / \chi_n^t(\Theta^t) A_n^t(T) \sqrt{k_n^t}\Big)$$

$$-\mathrm{Ln}\Big(I(T_o) S(T_o) \chi_o(\Theta)\, A_o(T_o)\sqrt{k_o} / I^t(T_o) S^t(T_o) / \chi_o^t(\Theta^t) A_o^t(T_o) \sqrt{k_o^t}\Big).$$

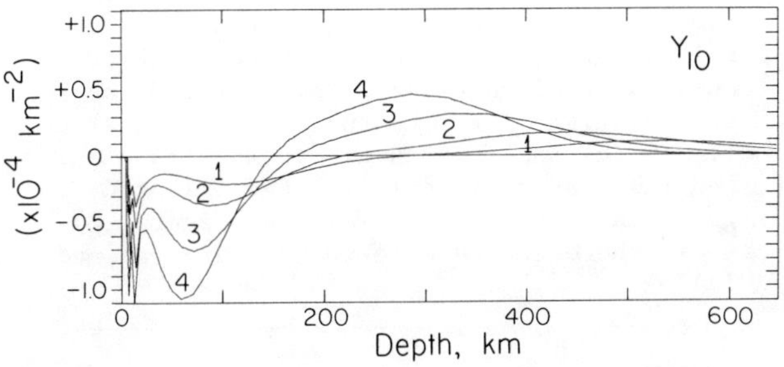

Fig. 1. Partial derivative Y_{10} of $\delta_{10}(T_o)$ with respect to Q_s^{-1}: (1), T_o= 100 sec.; (2), T_o= 80 sec.; (3), T_o= 60 sec. and (4), T_o= 50 sec.; The curves are computed for the starting model of ocean used by Cara (1979) and are normalized to unit layer thicknesses.

Of particular interest is the coefficient $\delta\gamma_{no}(T_o)$ at it does not depend on the source time function and on the instrumental response. The method proposed here may thus work well even if the product I(T)S(T) is unknown, provided that the geometrical source parameters are known.

Now, we have to address the following question: what is the geophysical interest of a "datum" $\delta\gamma_{no}(T)$ or $\delta\gamma_{no}(T_o)$ since we cannot infer the apparent quality factor $Q_n(T)$ or $Q_n(T_o)$ of each individual mode n from such data? The situation is quite similar to that for phase and group velocities when looking for an elastic model from dispersion measurements. Indeed it is simpler. The attenuation coefficients $\gamma_n(T)$ are related to the apparent quality factor $Q_n(T)$ of the mode n by the relation (Brune, 1962):

$$\gamma_n(T) = \pi/u_n(T)\ Q_n(T)\ T,$$

where $u_n(T)$ is the group velocity. $Q_n^{-1}(T)$ can in turn be linearly related to the intrinsic internal friction in shear Q_s^{-1} and compression Q_k^{-1} for small dissipation (Anderson and Archambeau, 1964). The relative attenuation coefficients $\delta\gamma_{no}(T_o)$ may thus be linearly related to the intrinsic internal friction $Q_s^{-1}$and Q_k^{-1}at the period T_o. Figure 1 shows the partial derivatives of such coefficients with respect to Q_s^{-1} at several periods. The curves are normalized to unit layer thicknesses. An important property of these partial derivative curves is that their integral over the whole depth range is very close to zero. So that if Q_s were independent of the depth at a given period T_o, one should expect a very small value of the "observed" relative attenuation coefficient $\delta\gamma_{10}(T_o)$.

Under the assumption of frequency independence of Q_s in the period range under study, the partial derivatives of the coefficients $\delta\gamma_{no}(T)$ have been computed for modes 0 (fundamental), 1 (first higher) and 2, (second higher), by choosing a reference period T_o equal to 100s.

Figure 2 shows that at a period of 30s, the coefficient $\delta\gamma_{10}$ will yield good estimates of the average Q_s between 100 and 300 km.

Data Analysis

The relative attenuation coefficients $\delta\gamma_{no}(T)$ are estimated in this section from four sets of WWSSN records obtained in the continental United States for intermediate-depth earthquakes located in the New Hebrides. Phase velocities have been already estimated from the same sets of records in a previous study (Cara, 1978). The coefficients $\delta\gamma_{no}(T)$ we obtain in this section are mainly related to the Pacific ocean, over the full path between the epicenters and the array of stations.

a) Data processing

The phase velocity curves obtained from the global analysis of the records within the array of stations in the previous study of Cara (1978) are used here as input data for the application

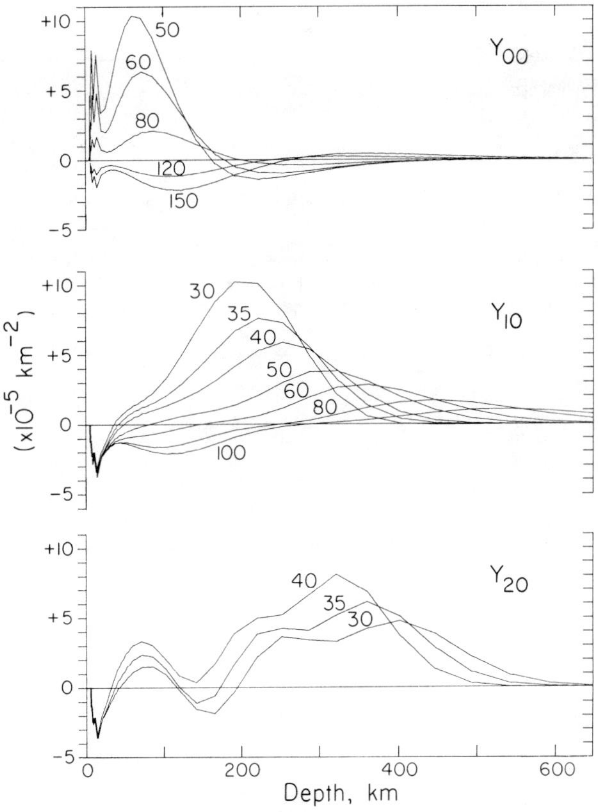

Fig. 2. Partial derivatives Y_{10} of $\delta\gamma_{10}(T)$ with respect to Q_s^{-1} for the different periods T marked on each curve in sec.. A frequency independence of Q_s is assumed in the computations. The curves are computed for the same model as in Figure 1 and are normalized to unit layer thicknesses.

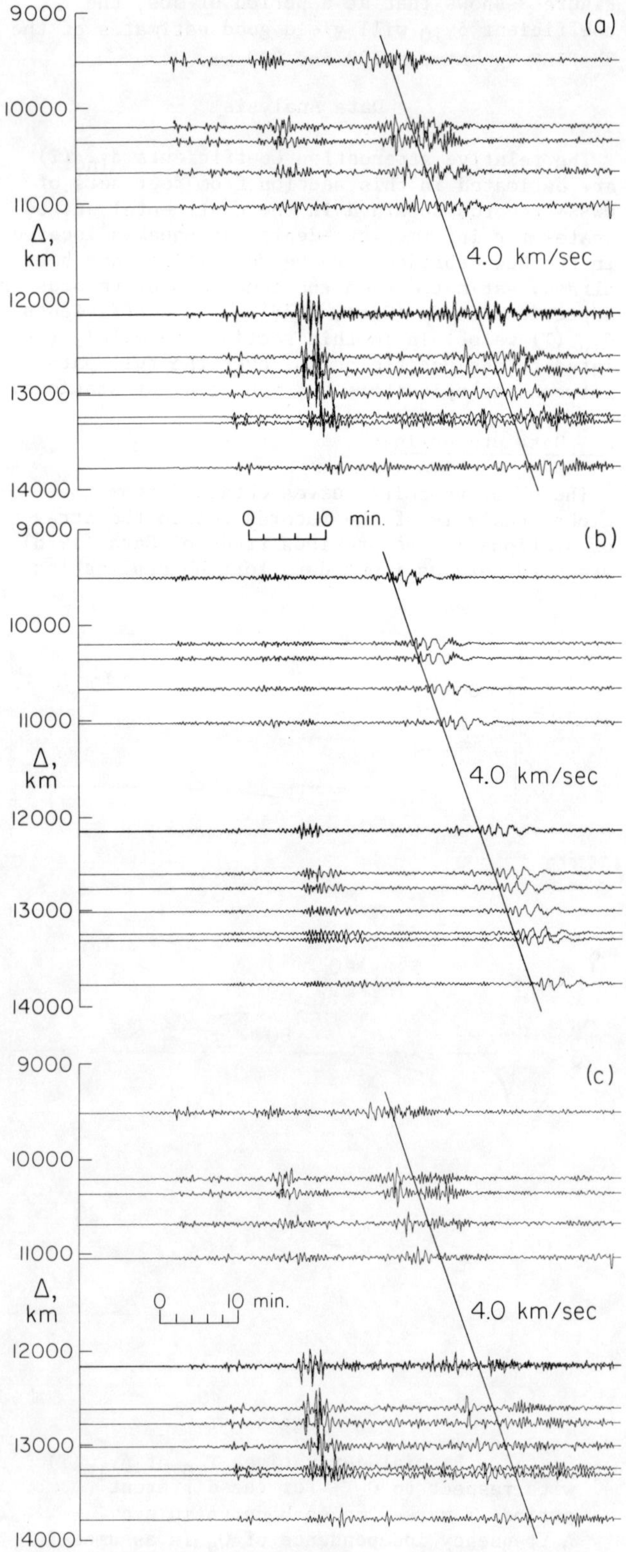

Fig. 3. Illustration of the spatial filtering technique of Cara (1978) for the event NH4 recorded across the United States (vertical component). The spatial Gaussian filters are centers on the phase-velocity curve of the fundamental Rayleigh mode and the relative filter bandwidth P_k is equal to 1. In (a) the records are shown versus epicentral distance and in (b) the filtered fundamental mode is shown. The figure (c) shows the difference between figure (a) and (b).

of the spatial filtering technique. Figure 3 shows an example of application of the spatial filters to the extraction of the fundamental mode. When applying this technique, care must be taken in choosing the bandwidth of the filters: too narrow filters yield an apparent attenuation of the waves which is sensitive to lateral variations of dispersion within the array; too wide filters on the other hand do not allow us to significantly improve the signal to noise ratio of the filtered mode. As in Cara (1978), a filter bandwidth parameter $P_k = 1.$ seems to be a good compromise.

Once spatial filters have been applied to the records, the amplitude level of the filtered mode is well improved, with respect to the other modes and the noise, but undesired signals can remain; Figure 3b shows for example that a slight residual higher mode signal remains in front of the filtered fundamental mode. As we shall see, such residual signals may affect the measurement of the fundamental mode amplitude of short period. To deal with these residual signals, it is necessary to use a time-frequency analysis of the filtered records instead of using a simple time Fourier transform. Several techniques may be employed for such a purpose: simple time-windowing of the records, time variable filtering and multiple frequency filtering (Dziewonski et al., 1969). The latter technique allows us to directly analyze the signal in the time-frequency plane and any secondary signals may be visualized. For that reason, I have chosen to use the multiple frequency technique in measuring the amplitude spectrum. By picking the maximum amplitude for each period on the time-frequency amplitude diagram, one obtains only a smooth estimate of the actual spectrum. To apply the method proposed in the previous section, it is necessary to estimate the theoretical spectrum through the same method of analysis.

The theoreticalamplitude spectrum have thus been estimated by passing synthetic seismograms computed for the source parameters given in Table I through the same frequency filters as for the actual records. To compute realistic seismograms, the instrumental response was taken into account and a very simple assumption has been made for the attenuation of the waves along the propagation paths: a constant apparent Q equal to 100 for each mode (Figure 4 shows that it is a reasonable choice).

TABLE I. Source parameters used in this study and estimation of the seismic moment from this paper (M is computed by assuming Q = 89.3 at 100s for the fundamental Rayleigh mode and M^* is computed for Q = 100). All the geometrical parameters are from Pascal et al. (1978) with the exception of δ, λ and θ of event NH1 which are from Isacks and Molnar (1971).

Event	NH1	NH2	NH3	NH4
Date	May 1, 1963	Jan. 19, 1969	Oct. 13, 1969	April 20, 1970
Latitude	19.0S	14.9S	18.8S	18.8S
Longitude	168.9E	167.3E	169.3E	169.3E
h	151 km	110 km	240 km	240 km
δ	41°	88°	--	78°
λ	24°	58°	--	90°
θ	-117°	-169°	--	98°
M	2.9 10^{26} d.cm	1.8 10^{27} d.cm	1.3 10^{26} d.cm	1.0 10^{26} d.cm
M^*	2.6 10^{26} d.cm	1.6 10^{27} d.cm	1.1 10^{26} d.cm	0.9 10^{26} d.cm

The synthetic seismograms have been computed for a flat layered elastic model by using a program written by Harkrider and Archambeau (1980). A complete comparison between observed and synthetic seismograms would imply the use of spherical computations, but due to the large error in amplitude measurements such computations are probably not necessary for our purpose that is to estimate the shape of the source amplitude spectrum.

The theoretical amplitudes shown in Figures 4, 6, 7 and 8 are computed for a point double-couple source model, a source time function approximated by a trapezoid of total duration 1.5s and model 1 shown in Figure 5. The geometrical source parameters are given in Table I. For the azimuth of the strike direction Θ with respect to the station of observation, the average azimuth toward the stations of the array has been used. For event NH3, as in Cara (1978) the same focal mechanism as for event NH4 has been used. The data points shown in the Figures 4, 6, 7 and 8 are the basic data used to infer the relative attenuation coefficients.

b) Error due to source parameters

Model. The sensitivity of the theoretical amplitude spectrum to changes in the elastic models used in the computations is the first question to be addressed. The effect of changing the elastic model is shown in Table IIa.

Model 1 is a homogeneous layer approximation of the elastic Pacific model obtained by Cara (1978) from the same sets of records. Model 2 exhibits a less pronounced low velocity zone, the S velocity distribution with depth being close to the average model of the Earth 1066A (Gilbert and Dziewonski, 1975). Both models 1 and 2 are oceanic models and may be regarded as reasonable candidates for the source region of

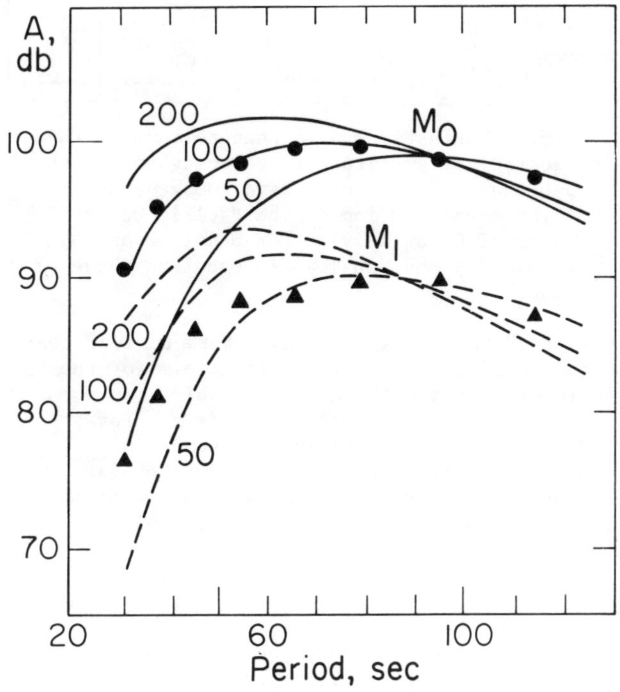

Fig. 4. Observed and theoretical amplitudes for the modes 0 and 1 excited by event NH1 and observed in the station BKS (after both spatial and multiple frequency filtering). Three constant apparent Q values have been tried to compute the theoretical amplitudes (Q = 50,100, and Q = 200). The amplitudes are given in decibel versus the period T, the value 100 db corresponding to the maximum of the theoretical amplitude for the mode 0.

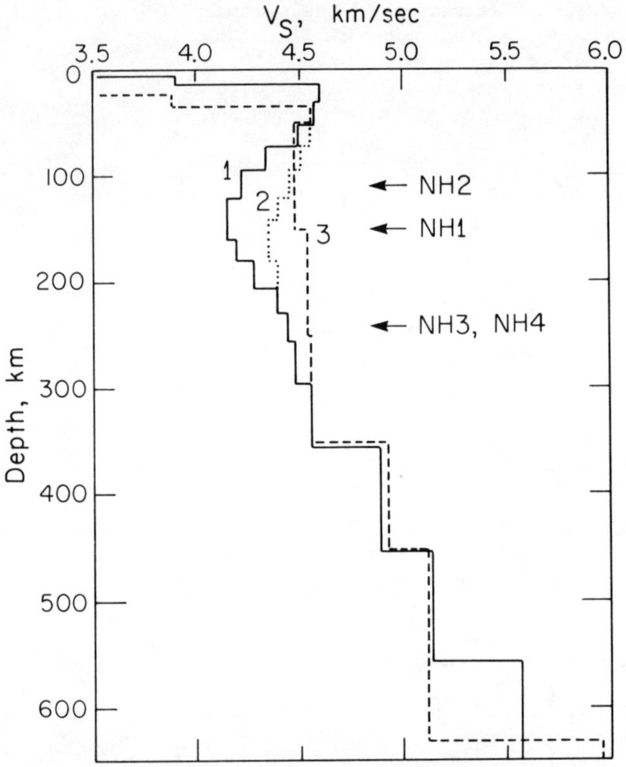

Fig. 5. Elastic models used to compute the theoretical amplitude (S velocity versus the depth h). The model 1 is a homogeneous layer approximation of the Pacific Ocean model of Cara (1979). For models 2 and 3, P velocity and density are equal to those of model 1.

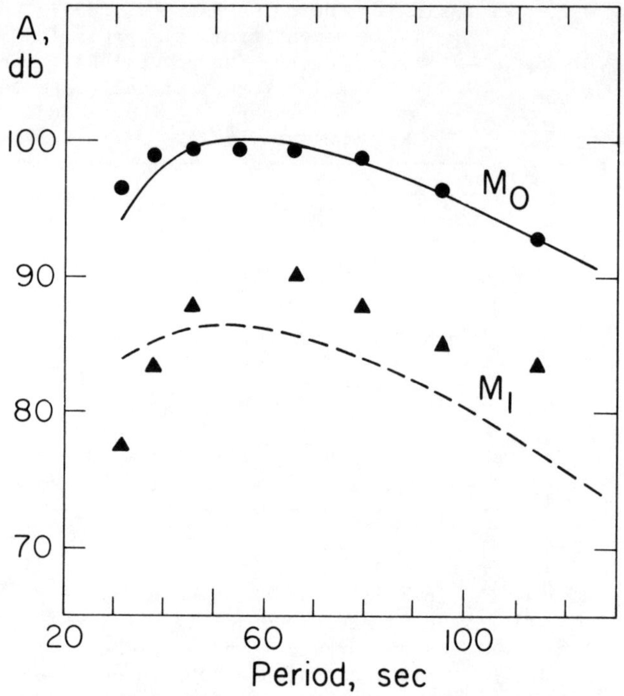

Fig. 6. The same as Figure 4 but for event NH2 and a constant apparent Q = 100.

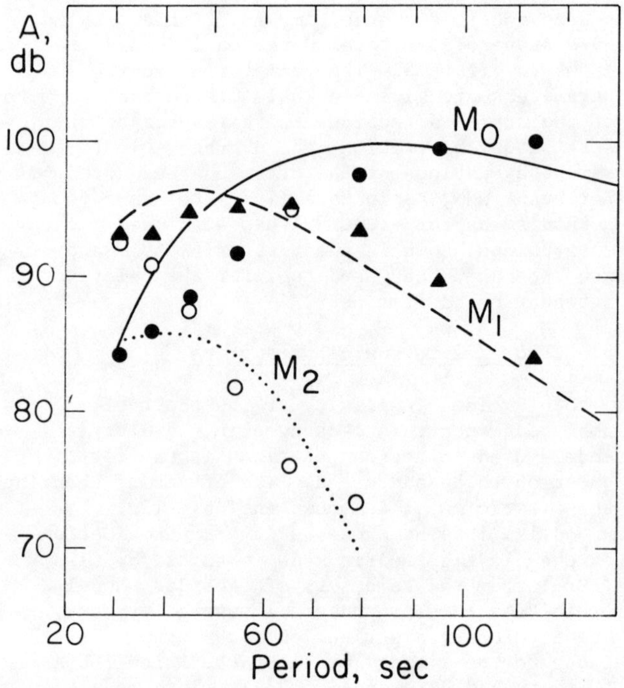

Fig. 7. The same as Figure 6 but for event NH3.

the New Hebrides earthquakes. Note however that the idea of choosing different models for the source region and the area over which the propagation of the waves occurs (e.g., Forsyth, 1975), supposes implicitly that negligible changes in amplitude occurs at the "boundary" between the two regions. As all the problems involving normal mode theory in laterally heterogeneous medium, such an assumption is difficult to justify due to the possibility of mode conversion. In this paper, such phenomena as well as the effects of scattering or lateral refractions are assumed to be negligible.

The perturbation $B_n(T)$ of the amplitude spectrum at the source for a mode n and a period T is given in Table II. These perturbations may be easily converted in terms of perturbations $\Delta\delta\gamma_{no}(T)$ of the relative attenuation coefficients defined in the previous section. From (5), we get

$$\Delta\delta\gamma_{no}(T) = \Delta\eta_{no}(T) = \frac{1}{x}\frac{\text{Ln } 10}{20}$$

Converted in terms of $\delta\gamma_{no}$ perturbations, the effect of replacing the model 1 by the model 2 (Figure 5) in the source region is less than

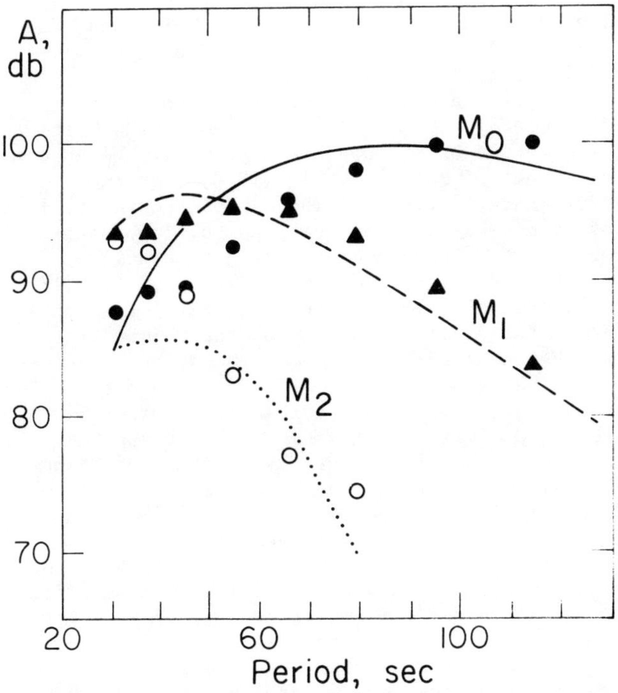

Fig. 8. The same as Figure 6 but for event NH4.

$0.13\ 10^{-4}\ km^{-1}$ for the fundamental mode at the smallest epicentral distance and $0.20\ 10^{-4}\ km^{-1}$ for the mode 1.

The effect of replacing the model 1 by a continental model (model 3 of Figure 5, derived from a Northern Eurasian model of Cara et al., 1980) is also shown in Table II. This gives an approximate upper bound on expected changes in amplitude due to change in the elastic model.

Source time functions. The theoretical amplitudes shown in Figures 4 and 6 to 8 are computed for a trapezoidal time source function of total duration 3 x 0.5 s. This time function does include both the finite rise time of the dislocation and its propagation effect over the fault plane. Accounting for this propagation effect by changing the source time function is valid for earthquakes with small source dimensions only.

We shall see that the seismic moments of the events used in this study vary from about $9\ 10^{25}$ dyne-cm to $6\ 10^{26}$ dyne cm. A recent study of 17 intermediate depth and deep earthquakes of the Tonga-Kermadec trench (Chung and Kanamori, 1979) has shown that for seismic moment in the range $1.8\ 10^{25}$ to $3.3\ 10^{26}$ dyne-cm the maximum dimensions of the fault are expected to be smaller than 24 km for all the intermediate depth events and the apparent source time duration is less than 6-8s for P-waves, depending on the model

TABLE IIa. Perturbation of the source amplitude spectrum of event HN1 (in db) due to changes in the source parameters given in Table I and changes from elastic model 1 in Fig. 5 (M1) to the models 2 (M2) and 3 (M3) shown in this figure. The effect of increasing the time duration τ to 12s is shown in the last column. The perturbations are computed relatively to the mode 0 at 100s the last line (*) giving the absolute perturbation for this mode at 100s.

	T (s)	$h+20_{km}$	λ+15°	δ+15°	θ+15°	M2	M3	τ = 12s
Mode 0	120	0.4	0.0	-0.3	0.2	0.0	-0.5	0.1
	100	0.0	0.0	0.0	0.0	0.0	0.0	0.0
	80	-0.5	0.1	0.3	-0.1	0.2	0.6	0.0
	60	-1.1	0.1	0.4	-0.1	0.6	1.4	-0.1
	50	-1.5	0.1	0.5	-0.1	0.8	1.6	-0.2
	40	-2.0	0.0	0.4	-0.1	0.9	1.2	-0.4
	30	-2.7	0.0	0.3	-0.1	1.1	-0.9	-0.9
	(*)	-0.4	-1.1	-2.2	1.6	-1.4	-2.0	-0.2
Mode 1	100	0.6	1.0	2.7	-1.5	-0.6	-0.5	
	85	0.3	1.0	2.7	-1.5	-0.4	-0.6	
	70	0.0	0.9	2.6	-1.4	-0.2	-0.7	
	55	-0.4	0.8	2.4	-1.2	-0.1	-0.8	
	40	-0.5	0.2	1.0	-0.3	+0.7	-1.5	
	32.5	0.0	-0.6	-1.3	0.8	+1.7	-1.6	

TABLE IIb. The same as Table IIa but for event NH4.

	T	h+20 km	λ+15°	δ-15°	δ+12°	θ+15°
Mode 0	120	0.3	0.0	0.0	0.0	0.0
	100	0.0	0.0	0.0	0.0	0.0
	80	-0.4	0.1	0.0	-0.1	0.0
	60	-1.0	0.0	0.0	-0.1	-0.1
	50	-1.5	0.1	-0.1	0.0	0.0
	40	-2.0	0.1	-0.2	-0.7	0.0
	30	-3.2	-0.2	-0.7	-0.3	-0.4
	(*)	-1.1	-0.3	-0.9	0.3	-0.6
Mode 1	100	1.3	0.5	5.3	-4.1	0.0
	85	1.7	0.2	3.0	-1.3	0.0
	70	1.7	0.0	-0.4	0.1	0.0
	55	1.4	-0.3	-3.5	0.5	0.0
	40	0.6	-0.1	-3.5	0.5	0.0
	32.5	0.0	-0.1	-2.3	0.4	0.0
Mode 2	41	-1.0	0.2	6.3	-13.5	-0.4
	34	4.1	1.3	6.4	-4.3	0.7
	30	4.7	1.5	4.1	0.1	1.3

used for the fault propagation. Due to slower phase velocity for Rayleigh waves than for P-waves, the apparent far field source time duration may be greater for the data used in this paper. The effect of increasing the source time duration from 1.5s to 12s is shown in Table II. Such an effect is negligible as compared to changes in other source parameters for periods longer than 50s and remain smaller than 1 db at 30s. This strongly supports the validity of using a point double couple source with a small time duration in modeling the excitation of the modes.

Focal depth. The effect of increasing source depth 20 km is shown in Table IIa for event NH1 and in Table IIb for event NH4. Such perturbations may change $\delta\gamma_{00}$ by 0.39 10^{-4} km^{-1} and $\delta\gamma_{20}$ by 0.54 10^{-4} km^{-1} at 32.4s for event NH4 if the reference period is 100s. On the other hand, the focal depth is a less critical parameter for the coefficient $\delta\gamma_{10}$ of both events. The great sensitivity of $\delta\gamma_{20}$ to the focal depth is very unfortunate since important geophysical information should be available from this mode which is fairly well observed for event NH4 between 30 and 40s (see Figs. 7 and 8).

Orientation of the double couple. The three geometrical source parameters, Θ, λ, and δ have been changed by 15° for events NH1 and NH4. Such changes in the angular parameters may be considered as generous upper bounds when inspecting the focal mechanism solutions of Isacks and Molnar (1971) for event NH1 and Pascal et al. (1978) for event NH4.

Recently, a new focal mechanism has been determined by Chung (1979) for event NH4. The orientation of the double couple is in agreement with that of Pascal et al. (1978) (maximum difference 5° for λ). The dip of the fault plane of event NH4 might be the cause of important errors if this parameter were not so well constrained by the body wave data (see Fig. 16 of Pascal et al., 1978). Uncertainties of about 1 db in the ratio of mode 1 to mode 0, due mainly to the parameter δ, is a more reasonable estimate than that given in Table IIb for this event. For event NH1, δ is also a critical parameter but seems to be less well constrained by the body wave data than for event NH4 (Isacks and Molnar, 1971). An error on δ and Θ should explain a general bias of about 0.3 10^{-4} km^{-1} for $\delta\gamma_{10}$ between 50 and 100s, for event NH1.

Event NH2 has been studied in detail by Chung and Kanamori (1978). Its focal mechanism differs in an appreciable manner from that of Pascal et al. (1978), mainly for the strike direction. Direct computation shows that the absolute amplitude of the mode 0 and 1 are changed by about +9.5 db whereas the relative amplitude of mode 1 to mode 0 is changed by less than 0.5 db in the period range under study. The uncertainty in the focal parameter of this event is thus important for the estimation of its seismic moment as it will be discussed in section c, but is not crucial for the estimation of the relative attenuation coefficients.

c) Experimental errors

The error due to digitizing the WWSSN photographic recored and the error due to the whole filtering procedure are probably very small. A possible source of bias due to the spatial filtering technique in presence of lateral heterogeneities is discussed below. On the other hand random errors in such data processing may be checked by comparing the results obtained for the two close events NH3 and NH4 of same focal mechanism and which are recorded on similar arrays of stations (Cara, 1978); Table III shows that the difference between both sets of measurements does not exceed 0.11 10^{-4} km^{-1} between 40 and 100s for the coefficients $\delta\gamma_{00}$ and $\delta\gamma_{10}$.

The effect of lateral variation of dispersion on the spatial filtering of the record may be checked on the fundamental mode. Event NH1 presents a well excited fundamental mode and the coefficients $\delta\gamma_{00}(T)$ may be estimated directly from the original records. The results are shown in Table III for both stations BKS and AAM; the values obtained after spatial filtering do not show systematic bias in the station AAM as compared with the "direct" measurements and show a slight bias toward higher attenuation for BKS. This means that the spatial filters used

TABLE III. $\delta\gamma_{no}(T) = \gamma_n(T) - \gamma_o(T_o)$ estimated for T_o = 100s (units in 10^{-4} km^{-1})

(1) Average of all data (with the exception of columns NH1*)

(2) Average of a limited set of data

* Direct measurements

		BKS					AAM					(1)		(2)	
	T(s)	NH1	NH1*	NH2	NH3	NH4	NH1	NH1*	NH2	NH3	NH4	$\delta\gamma_{no}$	$\Delta\delta\gamma_{no}$	$\delta\gamma_{no}$	$\Delta\delta\gamma_{no}$
Mode 0	32.4	1.42	1.36	1.33	(1.65)	(1.25)	0.97	0.84	1.30	(1.04)	(0.99)	1.21	0.25	1.20	0.24
	38.9	1.03	1.03	1.05	(1.64)	1.28	0.70	0.75	0.97	0.81	0.87	1.01	0.28	0.94	0.18
	46.7	0.77	0.71	0.87	1.46	1.36	0.58	0.65	0.70	0.70	0.76	0.86	0.30	0.86	0.30
	56.0	0.58	0.52	0.66	1.11	1.05	0.45	0.59	0.45	0.71	0.74	0.67	0.23	0.67	0.23
	67.2	0.35	0.29	0.42	0.73	0.66	0.34	0.38	0.30	0.59	0.58	0.46	0.16	0.46	0.16
	80.6	0.16	0.12	0.14	0.40	0.36	0.16	0.21	0.07	0.31	0.30	0.23	0.10	0.23	0.10
	96.7	0.00	0.00	0.00	0.00	0.00	0.00	0.00	0.00	0.00	0.00	0.00	0.00	0.00	0.00
	116.0	-0.26	-0.20	-0.10	-0.35	-0.34	-0.22	-0.17	-0.19	-0.33	-0.28	-0.24	0.08	-0.24	0.08
Mode 1	32.4	(1.95)	--	(2.30)	1.58	1.53	(1.15)	--	--	(1.64)	(1.84)	1.71	0.36	1.55	0.04
	38.9	(1.51)	--	(1.38)	1.40	1.36	(1.00)	--	--	1.30	1.37	1.35	0.16	1.36	0.04
	46.7	(1.10)	--	0.59	0.95	0.97	(0.92)	--	--	0.82	0.86	0.89	0.16	0.84	0.15
	56.0	0.87	--	--	0.54	0.52	0.76	--	--	0.43	0.47	0.60	0.18	0.60	0.18
	67.2	0.64	--	-0.24	0.09	0.09	0.50	--	--	0.08	0.14	0.18	0.23	0.18	0.29
	80.6	0.23	--	-0.34	-0.23	-0.19	0.24	--	--	-0.21	-0.10	-0.09	0.23	-0.09	0.23
	96.7	-0.25	--	(-0.53)	(-0.41)	-0.35	-0.16	--	--	-0.34	-0.28	-0.33	0.12	-0.30	0.09
	116.0	-0.42	--	(-1.02)	(-0.46)	(-0.40)	-0.34	--	--	(-0.29)	(-0.52)	-0.49	0.24	-0.38	0.06
Mode 2	32.4	--	--	--	0.54	0.48	--	--	--	0.60	0.71	0.58	0.10	0.58	0.10
	38.9	--	--	--	0.39	0.25	--	--	--	0.42	0.33	0.35	0.07	0.35	0.07
	46.7	--	--	--	0.48	0.33	--	--	--	0.01	0.23	0.26	0.20	0.26	0.70

to improve the signal to noise ratio of the fundamental mode present a wide enough bandwidth to avoid significant bias in the presence of the known strong lateral variations of phase velocity within the array of stations. As the same filter bandwidth is used for higher modes, this increases our confidence in the values reported for these modes.

Nevertheless, the use of rather wide filter bandwidth for the spatial filtering (P_k = 1, see Cara, 1978) does not allow us to avoid completely interference effects between the filtered mode and the other, undesired modes. Figures 9a to 9c show the values of $\delta\gamma_{no}(T)$ estimated from the amplitudes of event NH3 at two stations of the array: BKS and AAM. The difference in epicentral distance between the two stations is 3264 km. The coefficients $\delta\gamma_{10}(T)$ are very stable when passing from one station to the other but the coefficients $\delta\gamma_{00}(T)$ and $\delta\gamma_{20}(T)$ show strong instabilities. These instabilities occur when the observed amplitudes are respectively lower than -6 db and -10 db of the maximum observed amplitude (see Figure 7). This suggests an interference effect with the residual first higher Rayleigh mode. An independent convincing argument may be added. The phase shift $\delta\psi$ between the modes 1 and 2 when passing from BKS to AAM can be computed by using the phase velocities obtained across the whole United States (Cara, 1978); $\delta\psi$ is found to be nearly constant in the period range 40-70s and equal to about 1.5 cycle, a value which can explain a strong beat effect between the stations.

As the residual interference effect may play a major role in the experimental error the coefficients $\delta\gamma_{no}$ have been estimated at two stations of the array for each event. One could

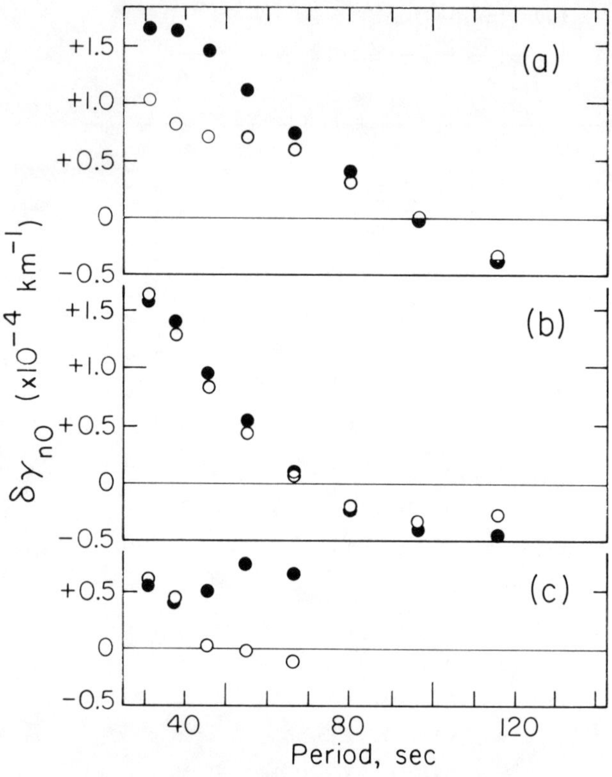

Fig. 9. Relative attenuation coefficient $\delta\gamma_{no} = \gamma_n(T) = \gamma_0(To)$ (To = 100 sec.,) for event NH3. The full circles are measurements made at the station of the array which is closest to the epicenter (BKS) and the open circles show the measurements made 3264 km farther (AAM).

furthermore expect that the scatter of the data may be reduced by eliminating the coefficients $\delta\gamma_{no}(T)$ estimated from the lowest amplitudes. For this purpose we have written in parenthesis the values inferred from amplitudes lower than -12 db of the maximum amplitude in Table III and two averages of the data are given: in column (1) all the data are taken into account whereas in column (2) the values in parenthesis are eliminated.

It must be noticed finally that the discrepancy between the measurements performed at the two stations AAM and BKS may reach 0.8 10^{-4} km^{-1} at 32.4s which is greater value than the error expected from uncertainty in the source parameters. On the other hand for $\delta\gamma_{10}$ at 96.7s this discrepancy is smaller than 0.1 10^{-4} km^{-1}, a smaller value than the error expected from source parameter uncertainties. It is difficult to say if the standard deviations reported in the column (1) of Table III are reasonable estimates of the standard error on the mean but the scatter observed for individual values in Table III may be explained by both the uncertainty in source parameters and by the above mentioned interference effect.

Results and Discussion

Before we examine in detail the results given in Table III, let us first estimate the coefficients $\delta\gamma_{10}(T_0)$ for different reference periods T_o. Such coefficients may be interpreted without any assumption upon the frequency dependence of Q within the Earth and they are not sensitive to uncertainties in the source time function and the instrumental response.

a) Results involving no assumption of the frequency dependence of Q

The values of the coefficients $\delta\gamma_{10}(T_0)$ are given in Table IV. Although these data exhibit a large scatter, $\delta\gamma_{10}(T_0)$ is clearly positive at periods smaller than 40s and is clearly negative beyond 80s. Due to the large standard deviation of the data, a zero value of $\delta\gamma_{10}(T)$ is expected somewhere between 40 and 80s.

If at a fixed period T_o, the intrinsic Q were independent of depth, one would expect a near zero value for $\delta\gamma_{10}(T_0)$ ($|\delta\gamma_{10}|<0.05 \times 10^{-4}$km^{-1} as shown by direct computations for $Q_s = 100$). The existence of negative values of $\delta\gamma_{10}(T_0)$ at long period seems well established. Events NH3 and NH4 exhibit for example an average $\delta\gamma_{10}$ of 0.53 10^{-4} km^{-1} at 80.6s (standard deviation 0.10 10^{-4} km^{-1}). Explaining such a value by an error in the source parameter would imply a bias of about -5 db in the theoretical amplitudes which seem difficult to achieve for these events.

By inspection of figure 1, one can conclude that a negative value of $\delta\gamma_{10}$ beyond 80s implies that Q increases with depth below 200 km. The zero-value of $\delta\gamma_{10}$ observed near 50s means that the balance of average internal friction Q^{-1} in the depth ranges 0-150 km and 150-700 km are roughly equal. Finally, positive values of $\delta\gamma_{10}$ near 30s are compatible with the existence of a higher Q in the upper part of the mantle. Such gross features are qualitatively in good agreement with the idea of a high Q lithosphere overlying a low Q asthenosphere with an increase of Q below 200 km.

b) Frequency-independent Q models

We now interpret fully the data listed in Table III by assuming that the frequency dependence of Q may be neglected in the period range 30-120s. We require furthermore that all losses in elastic energy should occur in shear. In the mantle, such an assumption is generally considered to be valid (Anderson et al., 1965). In this paper, the data are interpreted through a trial hypothesis method. Considering the large standard deviation exhibited by the data, only a few model parameters may be resolved and a more

TABLE IV. $\delta\gamma_{10}(T_0) = \gamma_1(T_0) - \gamma_0(T_0)$ given in 10^{-4} km^{-1}

	BKS			AAM			(1)		(2)	
T_0(s)	NH1	NH3	NH4	NH1	NH3	NH4	$\delta\gamma_{10}$	$\Delta\delta\gamma_{10}$	$\delta\gamma_{10}$	$\Delta\delta\gamma_{10}$
32.41	(0.53)	(-0.07)	(0.28)	(0.18)	(0.00)	(0.85)	0.39	0.34	--	--
38.89	(0.48)	(-0.24)	0.08	(0.30)	0.49	0.50	0.27	0.30	0.36	0.24
46.66	(0.33)	-0.51	-0.39	(0.34)	0.12	0.10	0.00	0.36	-0.17	0.33
55.98	0.29	-0.57	-0.53	0.31	-0.28	-0.27	-0.17	0.39	-0.17	0.39
67.17	0.29	-0.64	-0.56	0.17	-0.51	-0.44	-0.28	0.40	-0.28	0.40
80.60	0.07	-0.64	-0.55	0.08	-0.52	-0.40	-0.33	0.32	-0.33	0.32
96.70	-0.25	-0.41	-0.35	-0.16	-0.34	-0.28	-0.30	0.09	-0.30	0.09
116.03	-0.16	-0.11	(-0.07)	-0.12	(0.04)	(-0.24)	-0.11	0.09	-0.11	0.04

( ) One or two amplitudes used for the estimation of $\gamma_{10}(T_0)$ is smaller than -12 db of the maximum amplitude.

(1) Average on all values $\delta\gamma_{10}(T_0)$

(2) Average taken by excluding the values in parenthesis.

sophisticated inversion approach would probably yield similar results.

The figure 10 shows that the observed coefficients $\delta\gamma_{no}$ cannot be fitted by depth-independent Q_S models as already mentioned. However it must be noticed here that the constant Q_S values lying between 102 and 115 proposed by Sailor and Dziewonski (1978) for the depth range 0-670 km, are roughly compatible with the observed $\delta\gamma_{00}$ and $\delta\gamma_{10}$ within 0.3-0.4 10^{-4} km^{-1}, the largest discrepancy occurring for $\delta\gamma_{10}$ at periods greater than 80s. The coefficient $\delta\gamma_{20}$ is on the other hand very poorly predicted by such a depth-independent Q_S model but we know that large errors can affect the observed second higher mode amplitude.

Indeed, more can be done to improve the fit of the observed $\delta\gamma_{no}$ shown in Figure 10. After several trials, a tentative model MQ1 (Figure 11) has been obtained which fit the observations fairly well (Figure 12). Once again, the largest discrepancy occurs for the second higher mode but a possible explanation for this discrepancy may be proposed by deepening the source of both events NH3 and NH4 by 20 km: $\delta\gamma_{20}$ is increased by almost 0.5 10^{-4} km^{-1} at 32.4 sec whereas $\delta\gamma_{10}$ is increased only by 0.1 10^{-4} km^{-1} at 46.7s and the average value of $\delta\gamma_{10}$ given in Table III is increased by less than 0.1 x 10^{-4} km^{-1} on the whole period range. Since an uncertainty of 10-15 km is currently considered to be reasonable for the depth determination of such events, an error in the source depth might explain, at least partially, the observed discrepancy for $\delta\gamma_{20}$.

Several trials have shown that the relatively high Q below 250 km is fairly well constrained by the data. However, details in the model, like the rapid increase of Q between 180 and 250 km, are not well constrained. The model MQ2 (Figure 11) which exhibits a smooth increase of Q_S below 180 km gives a better average fit for $\delta\gamma_{10}$ at

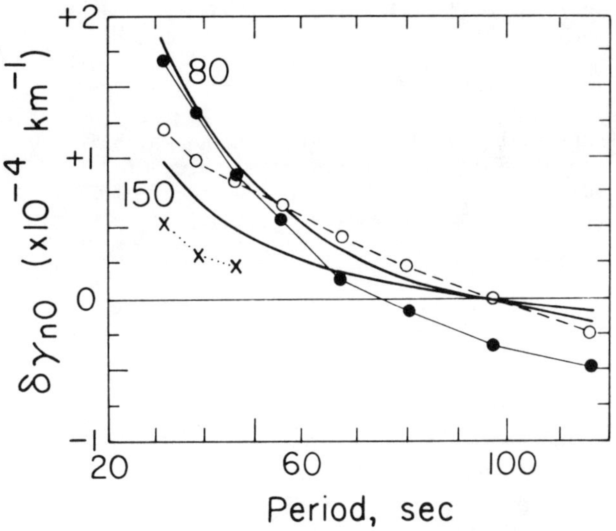

Fig. 10. Average observed values of $\delta\gamma_{no}(T)$ (Table III Column (2)) for the modes n = 0 (empty circles), n = 1 (full circles) and n = 2 (crosses). Theoretical curves for $\delta\gamma_{10}$ and depth independent Q_s models are shown by solid lines for Q_s = 80 and Q_s = 150 (the curves for $\delta\gamma_{10}$ are close to those for $\delta\gamma_{20}$ and $\delta\gamma_{00}$ within 0.1 10^{-4} km^{-1}).

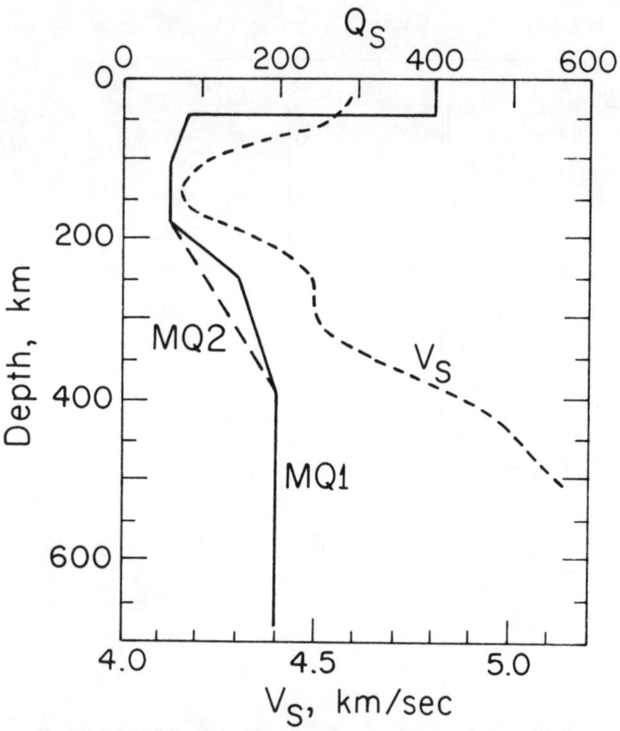

Fig. 11. Tentative frequency independent Q_s models to interpret the data shown in Fig. 10. The dotted line shows a S velocity model obtained for the Pacific Ocean by Cara (1979).

periods smaller than 60s but the discrepancy with the observed $\delta\gamma_{20}$ at 32.4s is increased (Figure 13). Indeed, as it stands, the observed value for $\delta\gamma_{20}$ at 32.4 s requires a very high value of Q_s between 300 and 400 km (see the partial derivatives in Figure 2.c) which is incompatible with the observed value of $\delta\gamma_{10}$ at 50s.

Considering the models of Figure 11 as fitting the data reasonably well, our purpose now is to compare these models with the results of other studies. In such a comparison two ranges of depth must be distinguished: (1) above 150 km, where several regional models have been obtained from surface wave observations; (2) between 150 and 700 km, where the only available models have been obtained from global normal mode data, eventually with constraints from body waves (Anderson and Hart, 1978a, 1978b).

TABLE V. Tentative Q_S models MQ1 and MQ2 of the Pacific Ocean (linear variation of Q_S is assumed between the given values).

Depth (km)	MQ1 Q_S	MQ2 Q_S
0	500	500
49	500	500
51	80	80
100	60	60
180	60	60
250	150	--
400	200	200
700	200	200

The models shown in Figure 11 are obtained for a region which is mainly oceanic over a path length greater than 9500 km but one must keep in mind that the data are an average of observations made at two stations in the continental United States after application of a stacking procedure to all the records observed in the array of stations. The low Q values observed in the upper mantle of Western United States (Solomon, 1972) and mode conversion across structural boundaries may thus affect these models.

However, the Pacific Ocean model obtained by Mitchell (1976) and the models MQ1 and MQ2 exhibit the same general features with a minimum value of Q_s near 150 km, but the first model shows higher Q_s values in the low Q zone (minimum value of about 80 instead of 60 here). Such a difference might be due to lateral variations of Q in the upper mantle: (1) the oceanic paths under study here cross a younger region on the average than the wide area studied by Mitchell et al. (1976) which includes the North Eastern Pacific (Canas and Mitchell have shown that lower values of Q may be expected in younger regions); (2) the effect of propagation across the array of stations may bias our results toward lower values of Q as suggested above.

Another discrepancy with the Pacific model of

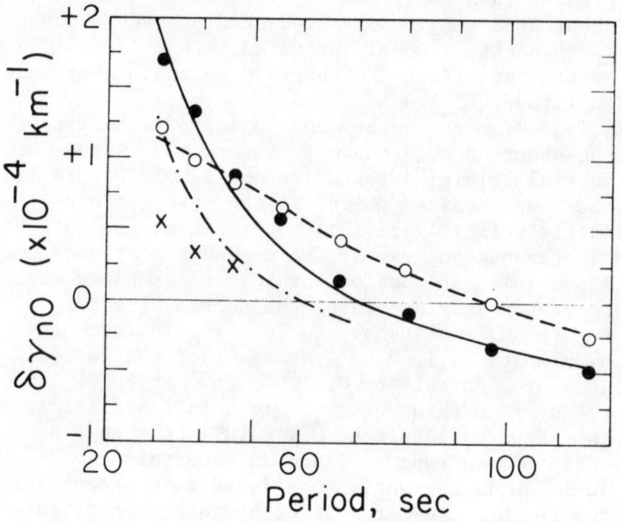

Fig. 12. The same as Figure 10 but the theoretical lines computed from the model MQ1 shown in Figure 11 (solid line, $\delta\gamma_{00}$; dashed line, $\delta\gamma_{10}$; dashes and points $\delta\gamma_{20}$).

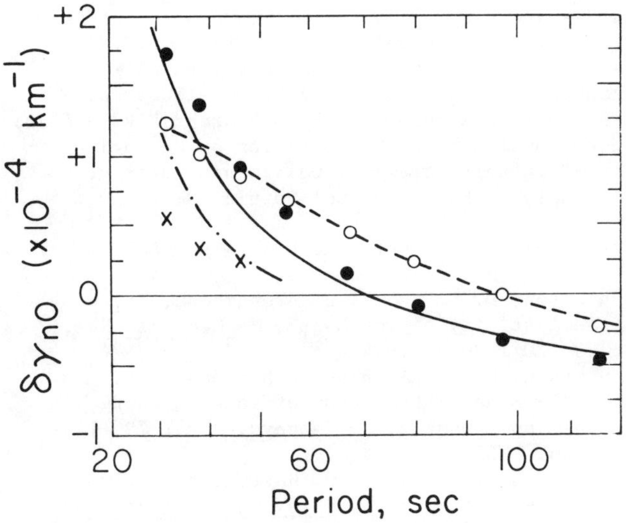

Fig. 13. The same as Figure 12 but for the model MQ2.

Mitchell (1976) appears when comparing the relative values of the attenuation coefficients predicted by this model (Figure 9 of Mitchell et al., 1977) with the relative values we observed (Table III). The very low values predicted for the first higher Rayleigh mode attenuation coefficient at long period is probably due to high values of Q_s set in this model, at depth greater than 300 km. It is then easy to explain such a discrepancy because the data used by Mitchell (1976) very poorly constrained Q_s below 150-200 km.

Below the depth of 150 km, models obtained from global normal mode data are more pertinent for a comparison purpose. It has been already mentioned that the models obtained by Sailor and Dziewonski (1978) exhibit lower Q_s values in the depth range 250-700 km than the models MQ1 or MQ2 shown in Figure 11. The models obtained by Anderson and Hart (1978a, 1978b) show higher Q_s in this depth range and are thus in better agreement with the models proposed here. The final model of Deschamps (1977) exhibits a nearly constant Q_s of about 200 between 400 and 700 km and is very close to the models MQ1 or MQ2 in this depth range.

Relatively high values of Q_s below 300-400 km are necessary to explain the high values of Q observed by several authors for the first higher Rayleigh mode in the period range 70-150s. The apparent Q predicted by the "regional" models MQ1 or MQ2 are however lower than those obtained by those authors: the model MQ1 predicts Q = 124 at 110s whereas Sailor and Dziewonski (1978) observed Q = 158 as a global average at 110.9s and in the range of period 105-115s, Jobert and Roult (1976) observed Q between 175 and 217; at 70s Okal observed Q = 140 and Q = 145 whereas the model MQ1 predicts Q = 115 at this period. If the pronounced low-Q-zone of the model MQ1 is a "regional" feature of a young oceanic area, it can bias significantly the apparent Q of the first higher Rayleigh mode toward lower values. Under such an assumption the Q_s values of the model MQ1 and MQ2 below 300 km are probably high enough to explain the high Q values found by the authors mentioned above as it is the case for example for the model SL1 of Anderson and Hart (1978) or the final model of Deschamps (1977).

Another interesting feature of the model MQ1 is its good correlation with the S velocity model inferred by Cara (1979) from the same sets of records as those used in this study, for the Pacific Ocean (Figure 11). The bottom of the low-Q-zone which is generally poorly constrained in seismological studies might thus coincide with the bottom of the low velocity zone.

c) Seismic moment

The seismic moment of the events used in this study can be estimated if an assumption can be made on the absolute value of the attenuation coefficient $\gamma_o(T_o)$ for the reference mode and the reference period T_o. A reasonable hypothesis for such an absolute value may be done on the basis of the models MQ1 or MQ2. At a period of 100s, the model MQ1 yields $\gamma_o = 0.84\ 10^{-4}$ km^{-1} (Q = 94.8) and the model MQ2 yields $\gamma_o = 0.89\ 10^{-4}$ km^{-1} (Q = 89.3) which is in good agreement with the average value of γ_o used by Mitchell (1976) at this period to investigate a Q_s model of the Pacific Ocean.

To show the dependence of the seismic moment upon assumption on apparent Q at 100s, two seismic moments are given in Table I, one computed for Q = 89.3 as in the model MQ2 and the other one with an arbitrary Q values equal to 100.

A striking fact from Table I is the large seismic moment of event NH2 as compared to the other events, of similar body wave magnitude. For event NH2, Chung and Kanamori (1978) had pointed out a discrepancy between the seismic moments inferred from body waves and surface waves. The seismic moment inferred by these authors from surface waves -5 x 10^{26} dynes.cm- is however about 3 times smaller than that reported in Table I. This discrepancy is due to the focal mechanism used in Table I which yields a node of excitation of Rayleigh waves rather close to our average direction of observation. By using the focal mechanism inferred from the radiation pattern of Rayleigh waves by Chung and Kanamori (1978) the values 1.8 and 1.6 10^{27} dynes.cm. in Table I are reduced to respectively 0.57 and 0.51 10^{27} dynes.cm, in very good agreement with the seismic moment obtained by these authors.

Conclusion

The method proposed here may be applied to infer Q models of the upper mantle related to the full path between an epicenter and a station. Only the shapes of both observed amplitude spectrum and theoretical source spectrum are needed to infer the relative attenuation coefficients which, in turn, may be linearly related to the intrinsic internal friction within the Earth if all other causes of attenuation (scattering, mode conversion, lateral refractions ...) may be neglected. To perform higher mode amplitude measurement in the period range 30-100s, the spatial filtering of the records observed on wide arrays of long period stations is probably the best way to deal with interference effects between modes which can be a great source of errors. However, when the method is used for the fundamental mode alone, single epicenter-station measurements are possible. The only, but serious, limitation is that the shape of the amplitude spectrum at the source must be very well constrained by other seismological data. Sources of small spatial extent but of high enough magnitude to be observed world-wide are good candidates for the application of this method by using a point double couple source model. Furthermore, intermediate-depth earthquakes may have a well constrained focal depth and are necessary to study the higher modes.

The results obtained in this study for the propagation of the fundamental and first higher Rayleigh modes across the Pacific Ocean and the array of WWSSN stations within the continental United States are consistent with a frequency-independent Q_s model exhibiting a low Q_s zone between 100 and 180 km ($Q_s = 60$) and an increase of Q_s below 200 km with an average Q_s value of 200 between 400 and 700 km. The results obtained for the second higher Rayleigh mode are poorly consistent with these models but the error analysis shows that the data may be in large error.

The increase of Q below the low Q zone of the upper mantle, initially proposed to occur below 400 km by Anderson and Archambeau (1964) seems thus to occur at shallower depth. The shape of this low Q zone region might then be well correlated with the shape of the low velocity zone as it is suggested by a comparison between the Q models of this paper and the S velocity model obtained by Cara (1979) for the Pacific Ocean.

Above the depth of 150-200 km, the model proposed here exhibits the same general features as the model proposed by Mitchell (1976) for the Pacific Ocean, although the values of Q_s found by this author between 100-200 km are higher than those proposed here.

Acknowledgments. I thank D. G. Harkrider who made his computer programs available to me and T. Lay for his help in the preliminary stage of this work. Thanks are also due to D. L. Anderson and H. Kanamori for helpful discussions and encouragements, W. Y. Chung who gave us a focal mechanism solution prior to publication and J. B. Minster who carefully read the manuscript. During the course of this work, the author was a Research Fellow of the California Institute of Technology and received funds from a C.N.R.S. - N.S.F. post-doctoral fellowship (French - U.S. convention) and from a N.A.T.O. fellowship. This work was supported by NSF Grant EAR77-14675. Contribution No. 3402, Division of Geological and Planetary Sciences, California Institute of Technology, Pasadena, California 91125.

References

Anderson, D. L. and C. B. Archambeau, The anelasticity of the Earth, J. Geophys. Res., 69, 2071-2084, 1964.

Anderson, D. L., A. Ben-Menahem and C. B. Archambeau, Attenuation of seismic energy in the upper mantle, J. Geophys. Res., 70, 1441-1448, 1965.

Anderson, D. L. and R. S. Hart, Attenuation models of the Earth, Phys. Earth Planet. Int., 16, 289-306, 1978a.

Anderson, D. L. and R. S. Hart, Q of the Earth, J. Geophys. Res., 83, 5869-5882, 1978b.

Anderson, D. L. and J. B. Minster, The frequency dependence of Q in the Earth and implications for mantle rheology and Chandler wobble, Geophys. J.R. astr. Soc., 58, 431-440, 1979.

Ben-Menahem, A. and D. G. Harkrider, Radiation pattern of seismic surface waves from buried dipolar point sources in a flat stratified Earth, J. Geophys. Res., 69, 2605-2620, 1964.

Brune, N. J., Attenuation of dispersed wave-trains, Bull. Seism. Soc. Am., 52, 109-112, 1962.

Canas, J. A. and B. J. Mitchell, Lateral variation of surface-wave anelastic attenuation across the Pacific, Bull. Seism. Soc. Am., 68, 1637-1650, 1978.

Cara, M., Regional variations of higher Rayleigh-mode phase velocities: a spatial-filtering method, Geophys. J.R. astr. Soc., 54, 439-460, 1978.

Cara, M., Lateral variations of S-velocity in the upper mantle from higher Rayleigh modes, Geophys. J.R. astr. Soc., 57, 649-670, 1979.

Cara, M., A. Nercessian and G. Nolet, New inferences from higher mode data in Western Europe and Northern Eurasia, Geophys. J.R. astr. Soc., 61, 459-478.

Chung, W-Y. and H. Kanamori, Source parameters and stress drop of intermediate and deep locus earthquakes in the New Hebrides, EOS. trans. Am. Geophys. Un., 60, 878, 1979.

Chung, W-Y. and H. Kanamori, Subduction process of a fracture zone and aseismic ridges - Local mechanisms and source characteristics of the New Hebrides earthquake of 1969 January 19 and some related events, Geophy. J.R. astr. Soc., 54, 221-240, 1979.

Chung, W-Y. and H. Kanamori, Variation of seismic source parameters and stress drops within a descending slab and its implications in plate mechanics, submitted to Phys. Earth Planet. Int., 1980.

Deschamps, A., Inversion of the attenuation data of free oscillations of the Earth (fundamental and first higher modes) Geophys. J.R. astr. Soc., 50, 699-722, 1977.

Dziewonski, A., S. Bloch and M. Landisman, A technique for the analysis of transient seismic signals, Bull. Seism. Soc. Am., 59, 427-444, 1969.

Forsyth, D. W., A new method for the analysis of multi-mode surface wave disperson: application of Love wave propagation in the East Pacific, Bull. Seism. Soc. Am., 65, 323-342, 1975.

Gilbert, F. and A. M. Dziewonski, An application of normal mode theory to the retrieval of structure parameters and source mechanisms from seismic spectra, Phil. Trans. Roy. Soc. Lond., A, 278, 187-269, 1975.

Harkrider, D. G. and C. B. Archambeau, Theoretical Rayleigh and Love waves from explosion in prestressed source regions (to be submitted for publication), 1980.

Isacks, B. and P. Molnar, Distribution of stresses in the descending lithosphere from a global survey of focal-mechanism solutions of mantle earthquakes, Rev. Geophy., 9, 103-174, 1971.

Jobert, N. and G. Roult, Periods and damping of free oscillations observed in France after sixteen earthquakes, Geophys. J.R. astr. Soc., 45, 155-176, 1976.

Kanamori, H. and D. L. Anderson, Importance of physical dispersion in surface wave and free oscillation problems: review, Rev. Geophys. Space Phys., 15, 105-112, 1977.

Liu, H. P., D. L. Anderson and H. Kanamori, Velocity dispersion due to anelasticity: implications for seismology and mantle composition, Geophys. J.R. astr. Soc., 47, 41-58, 1976.

Mitchell, B. J., L. W. B. Leite, Y. K. Yu and Hermann, Attenuation of Love and Rayleigh waves across the Pacific at periods between 15 and 110 seconds, Bull. Seism. Soc. Am., 66, 1189-1202, 1976.

Mitchell, B. J., Anelasticity of the crust and upper mantle beneath the Pacific Ocean from the inversion of observed surface wave attenuation, Geophys. J.R. astr. Soc., 46, 521-533, 1976.

Mitchell, B. J., N. K. Yacoub and A. M. Correig, A summary of seismic surface wave attenuation and its regional variations across continent and oceans, in The Earth's Crust, A.G.U. Monograph 20, 405-425, 1977.

Okal, E., Higher-mode Rayleigh waves studied as individual seismic phases, Earth Planet. Sci. Lett., 43, 162-167, 1979.

Pascal, G., B. L. Isacks, M. Barazangui and J. Dubois, Precise relocations of earthquakes and seismotectonics of the New Hebrides Island Arcs, J. Geophys. Res., 83, 4957-4973, 1978.

Sailor, R. V. and A. Dziewonski, Measurements and interpretation of normal mode attenuation, Geophys. J.R. astr. Soc., 53, 559-582, 1978.

Solomon, S. C., On Q and seismic discrimination, Geophys. J.R. astr. Soc., 31, 163-177, 1972.

Tsai, Y. B. and K. Aki, Precise focal depth determination from amplitude spectra of surface waves, J. Geophys. Res., 75, 5729-5743, 1970.

Weidner, D. J., The effect of oceanic sediments on surface wave propagation, Bull. Seism. Soc. Am., 65, 1531-1552, 1975.

Q^{-1} MODELS FROM DATA SPACE INVERSION OF FUNDAMENTAL SPHEROIDAL MODE ATTENUATION MEASUREMENTS

Seth Stein,[1*] Joseph M. Mills, Jr.[2] and Robert J. Geller[1]

1 Department of Geophysics, Stanford University, Stanford, CA 94305

2 Earth Sciences Division, Lawrence Livermore National Laboratory, Livermore, CA 94550

Abstract. We assemble a dataset for the attenuation of fundamental spheroidal modes covering the period 50 to 3000 seconds. The dataset combines surface wave and normal mode data from three studies, each of which explicitly includes realistic error bounds. We have inverted these data, using a data space inversion method with ranking and winnowing. Beginning with a series of models with uniform attenuation throughout the mantle, we derive smoothly varying models which fit the data quite well. Two different schemes to minimize deviations between the models and Q^{-1} data yield substantially similar results. We compare our models with previous studies, and show that a broad range of models is consistent with the fundamental mode Q^{-1} data. This range of models delineates our present resolution of attenuation within the mantle.

Introduction

Modern digitally recording seismic stations, have made it feasible to collect and analyze data appropriate for quantitatively describing the average seismic wave attenuation within the earth. In this study we use such new Q^{-1} data to obtain models of the earth's attenuation. Records from the High Gain Long Period (HGLP) stations, Seismic Research Observations (SRO) and International Deployment of Accelerometer (IDA) provided most of the data that we have used. The high quality and broad dynamic ranges of these instruments have made it possible to measure attenuation of Rayleigh waves and spheroidal free oscillations from 50 to 3000 seconds. Several different approaches have been taken to span this broad range of periods. Traveling surface waves have been used to estimate attenuation and group velocity for periods between 50 and 300 seconds [Mills and Hales, 1977, 1978]. Time domain estimates of the rate of decay of unsplit free oscillations were used for fundamental spheroidal modes from ${}_0S_6$ to ${}_0S_{28}$ (962 to 275 seconds) [Geller and Stein, 1979], and matching of synthetic seismograms incorporating attenuation with data filtered was used for split modes ${}_0S_2$ to ${}_0S_5$ (3300 to 1100 seconds) [Stein and Geller, 1978a]. The smooth transitions between data from the three period ranges demonstrate the lack of any bias between the different measurement techniques. Error estimates for the attenuation at each period are also included in the inversion schemes which we have used. Such an integrated high quality data set covering this wide period range provides a major new tool for the study of attenuation as a function of depth within the earth.

In this paper we first discuss the Q^{-1} dataset in detail. We then outline the data space inversion method, and the Q^{-1} models we have obtained. We show that widely differing Q^{-1} models can match the fundamental mode Q^{-1} observations.

The Analysis of Surface Wave Attenuation -- 50 to 300 Seconds

Measurements of the group or phase velocities and attenuations of surface waves fall into two classes: (1) regional studies usually restricted to periods of less than 160 seconds [e.g., Solomon, 1971; Hamada, 1972; Madariaga and Aki, 1972; Mitchell, 1973; Mitchell et al., 1975], and, (2) studies of great circle average velocities or attenuation coefficients which, in general, are not extended to periods of less than 160 seconds [Kanamori, 1970; Dziewonski, 1970; Jobert and Roult, 1976; Mills and Hales, 1977; Sailor and Dziewonski, 1978; Nakanishi, 1978, 1979]. Instrumentation used by standard networks

* Present Address: Department of Geological Sciences, Northwestern University, Evanston, Illinois 60201

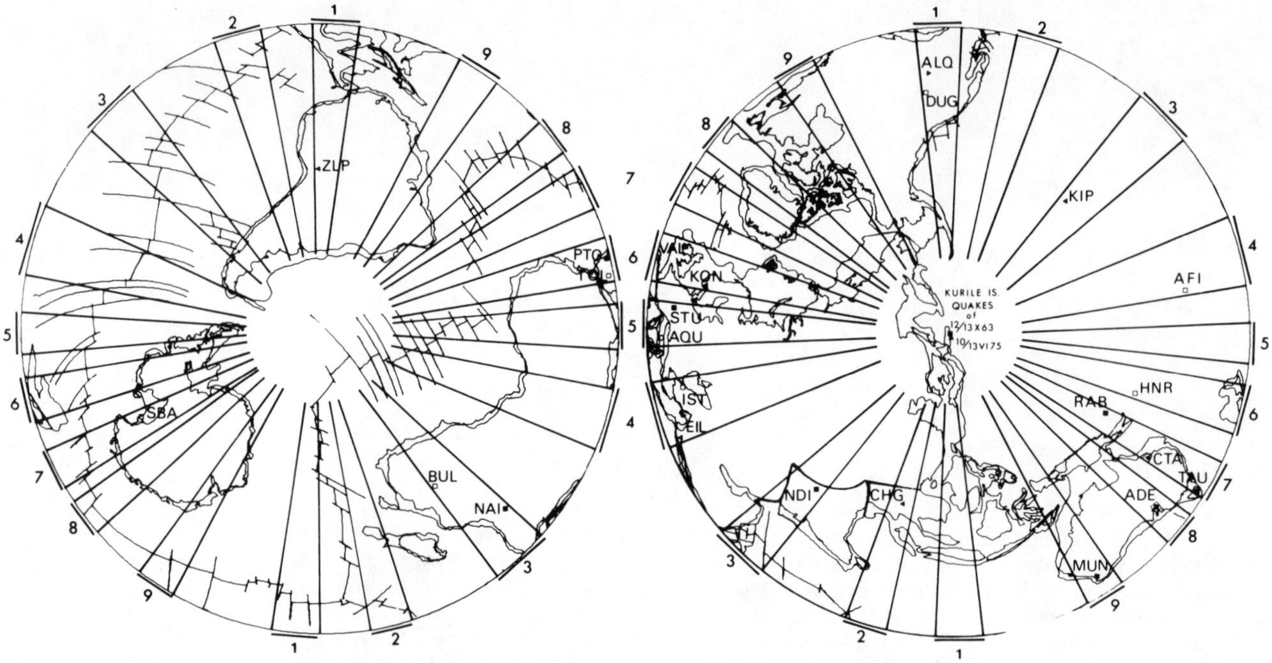

Fig. 1. Great Circle Paths - Equal area projections of two heimspheres of the earth are shown with the Kurile Islands earthquakes of 12 and 13 October 1963 (square) and 10 and 13 June 1975 (triangle) located at one pole. Triangles represent HGLP stations and squares represent WWSSN stations. Azimuth windows for 9 great circle paths are numbered in both hemispheres. Coast lines and continental margins are shown for all continents.

(WWSSN, HGLP, and SRO) will not adequately record ground motion of the first few passages of surface wave trains from earthquakes larger than magnitude 7. Few of these studies, if any, contain information about the same great circle paths.

Mills and Hales [1977] presented measurements of great circle Rayleigh wave group velocity and attenuation for periods between 150 and 600 seconds for 7 great circle paths based on data from the M_S=8.1 Kurile Islands earthquake of 13 October 1963 (Figure 1). Mills and Hales [1978] extended the range of these measurements to periods as short as 50 seconds, and added more data for periods between 150 and 250 seconds to those found previously for the same and two additional paths. These measurements from 3 smaller Kurile Islands earthquakes of M_S=6.6 to 7.1 (Figure 1) provide an even azimuthal coverage of the earth. The paths traverse all major structural features of the earth: stable continental platforms, ocean basins, mid-ocean ridge zones, and island arcs. The Himalayas are also well sampled. The global average group velocities and attenuation coefficients should therefore represent good areal averages for the whole earth. This is especially important since lateral heterogeneity strongly affects surface waves at periods shorter than 250 seconds.

The attenuation coefficient for period T is defined as:

$$\gamma_i = \frac{1}{C} \ln \quad (A_i / A_{i+2}) \qquad (1)$$

where C is the circumference of the great circle path and A_i and A_{i+2} are the spectral amplitudes for period T of the Rayleigh phases R_i and R_{i+2}. Maximum amplitudes of Rayleigh phases in narrow-band filtered seismograms (also used for the determination of group velocity) have been used as estimates of the spectral amplitude. The average attenuation for each great circle path shown in Figure 1 was calculated by averaging all measurements of attenuation coefficients for all stations within each azimuth window for overlapping 20-second period ranges. Global mean attenuation coefficients (Figure 2) were calculated by averaging mean path attenuation coefficients for all paths, for each 20-second averaging window, again giving unit weight to each path.

Great circle group velocities and uncertainties were calculated from each pair of group velocity estimates, since some multiple of a great circle travel time can be formed from the sum or difference of travel times for any two Rayleigh phases observed at a single station, e.g.,

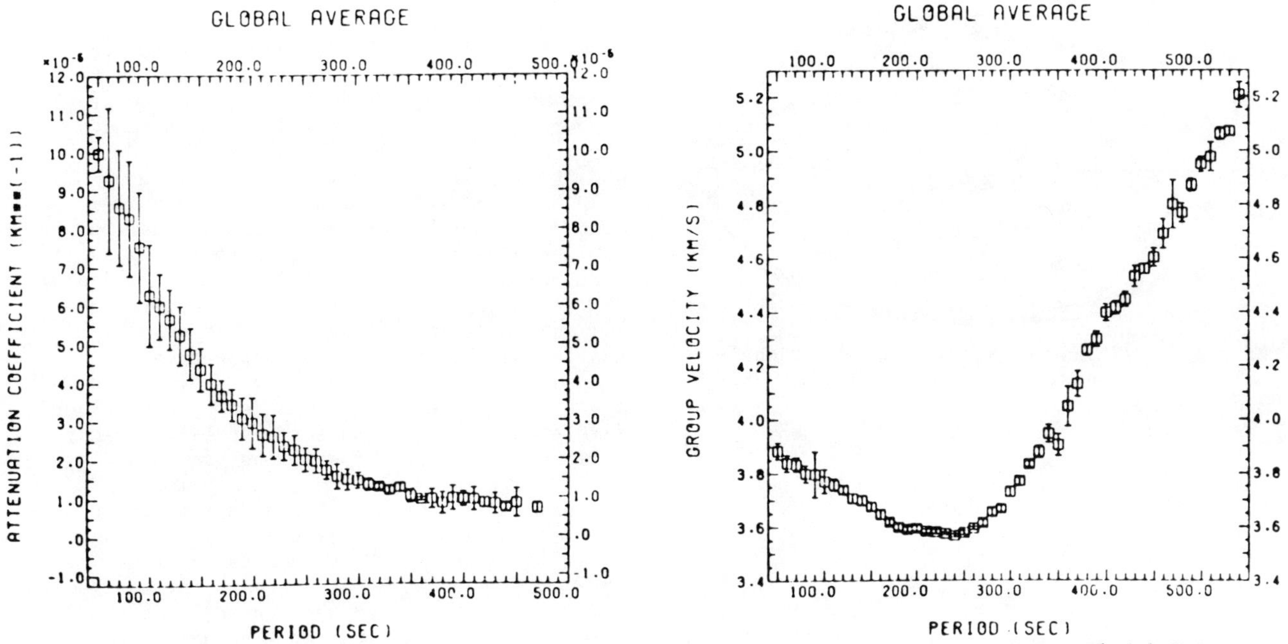

Fig. 2. Global Mean Great Circle Group Velocities and Attenuation Coefficients for Rayleigh Waves - have been calculated by calculating mean group velocities and attenuation coefficients for each of paths 1-9 (Figure 1), averaged over overlapping 20-second period windows and weighted equally for each path, to determine global mean group velocities and attenuation coefficients.

$$[t_1 + t_2] , \quad [t_3 - t_1] , \quad [t_2 + t_3] / 2 .$$

Means of all group velocity estimates for all stations on a great circle path (Figure 1) were formed for 20-second overlapping windows to produce averaged group velocity, as a function of period for each path. Global mean group velocities (Figure 2) were calculated by averaging individual path averages giving each path equal weight.

Attenuation coefficients of Mitchell et al. [1975] for Pacific Ocean paths for the period range 50 to 110 seconds, are within a factor of 2 of our global average estimates. Similar estimates for central United States paths by Mitchell [1973] and Herrmann and Mitchell [1975] indicate that attenuation coefficients at a period of 40 seconds are also within a factor of 2 of our estimate at 50 seconds. These global mean attenuation coefficients, $\gamma(T)$, were combined with global mean group velocites, $U(T)$, to find the specific attenuation at period T:

$$Q^{-1}(T) = \frac{1}{\pi} T \cdot U(T) \cdot \gamma(T) ,$$

which were used in the inversions discussed below.

Normal Mode Data

The data for periods longer than 300 seconds (${}_0S_{28}$) are taken from attenuation measurements for individual normal modes rather than from traveling surface waves. Two types of measurements are included here: (1) for the period range 962-275 seconds (${}_0S_6$-${}_0S_{28}$), the measurements are from unsplit normal modes; (2) for the four longer period modes ${}_0S_2$-${}_0S_5$, the measurements are from the split mode multiplets.

The unsplit mode measurements were made using ultra long period gravimeter records at seven stations of the IDA network [Agnew et al., 1976] for the August 19, 1977 (M_s 7 3/4) Indonesian earthquake. The IDA data provided the first opportunity to make attenuation measurements in this frequency band at multiple stations for a single earthquake. The location of the earthquake, and the seven stations used in the study are shown in figure 3.

Details of the analysis are given by Geller and Stein [1979]. Attenuation measurements were made in the time domain, by fitting a set of least squares lines to the envelope of the filtered time series. The results appear in the form shown in figure 4, showing an example for the modes ${}_0S_9$ and ${}_0S_{10}$. Each panel shows a measurement for a single mode at a given station. The data are the logarithm of the analytic signal, which represents the decay of the mode as a function of time.

Q^{-1} is measured by fitting straight lines to segments of the data ranging from the time of the peak amplitude until the time that the

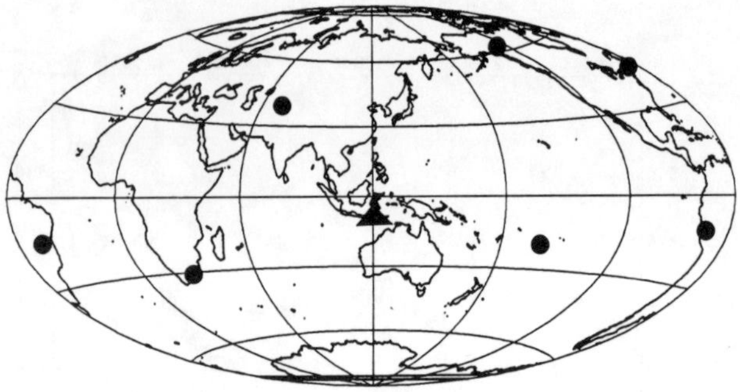

Fig. 3. 1977 Indonesian earthquake (triangle) and IDA stations (dots) used for Q^{-1} measurements on ${}_0S_6$ - ${}_0S_{28}$.

signal decays to e^{-1}, e^{-2} and e^{-3} of the peak respectively. Q^{-1} is then given by

$$Q^{-1} = -\bar{m}\ T/\pi,$$

where T is the period of the mode and $\bar{m}$ is the slope of the line. Thus, there are three Q^{-1} estimates for each mode, corresponding to the three least squares lines shown. These are given times 10^5, with errors derived from the standard least squares formula, using twice the number of frequency points in the passband as the number of degrees of freedom. (See Geller and Stein [1979] for details.)

This method was applied to the seven IDA records available for the Indonesian earthquake. Amplitude decay plots (like those in figure 4) were prepared for the twenty-three modes from ${}_0S_6$ to ${}_0S_{28}$. Plots of the envelope were made for each mode at each station, but only high quality data were retained in the final dataset. Data contaminated by either noise or nearby overtones were identified by a scalloped time series rather than a smooth linear decay.

Different numbers of stations had data of sufficiently high quality for each mode. In one case, ${}_0S_{11}$, none of the data was acceptable. Because the amplitude scale is logarithmic, the noise at the end of some records is of negligible amplitude. Some of the very low amplitude noise is the result of the zero fill at the ends of the records. In general, the data are ex-

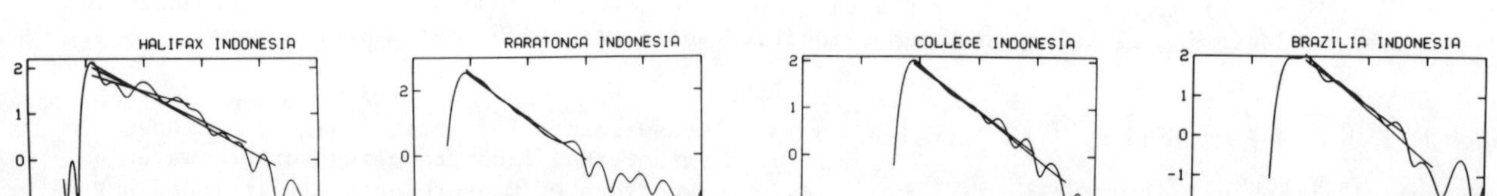

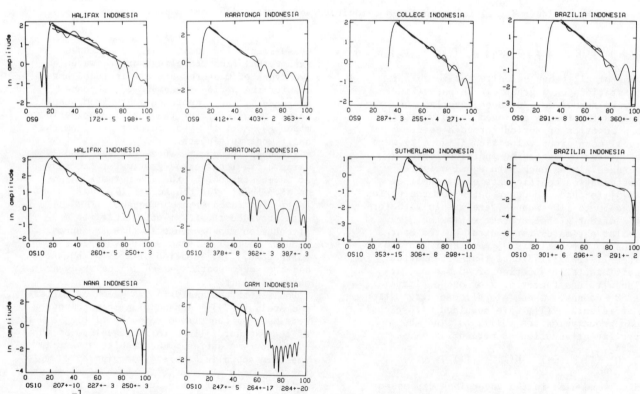

Fig. 4. Q^{-1} measurements for ${}_0S_9$ and ${}_0S_{10}$ from Indonesian earthquake records. Each box shows the logarithm of the unsmoothed envelope for a given mode at a single station, decaying as a function of time (the unit of the horizontal scale is hours after the origin). The vertical unit is natural logarithm of amplitude. The three numbers on the same line as the mode identification show $Q^{-1} \times 10^5$ and nominal error calculated drom the e^{-1}, e^{-2} and e^{-3} least squares lines. Note that the difference between stations often exceeds the nominal error bounds.

tremely good, and the values of Q^{-1} for each record are very well constrained.

The major advantages of time domain attenuation measurement are in its ability to minimize the effects of various types of noise. By studying the decay of the amplitude of a mode as a function of time, it is possible to directly determine the quality of the observations, in a manner that is not possible in the frequency domain. A damped harmonic oscillator, recorded with no noise, will show a standard resonance peak in the frequency domain or exponentially decaying amplitude in the time domain; its attenuation can be measured in either domain equally well. However, as noise is introduced, it is almost impossible to determine the amount of noise contamination in the frequency domain. On the other hand, the extent to which the signal in the time domain differs from a pure decaying exponential provides a straightforward way of demonstrating the quality of an observation. Noise bursts, resulting from an aftershock or some other source of contamination were easily detectable from the scalloping of the exponential decay; such biased observations were therefore rejected. (Computing the power in successive windows, an apparently equivalent method, is actually much less effective in detecting such effects.) Another advantage of time domain measurements is that they allow immediate identification of the ambient noise level, and signals at or below this level can be disregarded.

Examination of results at different stations for the same mode, e.g. ${}_0S_{10}$ at Raratonga and Brazilia, shows that the difference in Q^{-1} measured at different stations is much greater than the nominal error bound from the least squares fit. The least squares fit, then measures the uncertainty in the single station measurement, rather than that between stations.

We attribute the additional scatter to lateral heterogeneity, which is one cause of mode splitting [Madariaga, 1971; Usami, 1971; Saito, 1972; Luh, 1973, 1974]. Although individual singlets are not resolvable in this frequency band, a bias is introduced in the attenuation measurements [Buland, 1979]. This effect is analyzed in detail by Sleep, Geller and Stein [1981] who used the scatter to estimate lateral velocity and density heterogeneity. Any lateral heterogeneity of Q^{-1}, as opposed to elastic constants, may be a much smaller effect.

In compiling the attenuation data, we first took the mean of the Q^{-1} values at each station (obtained from the e^{-1}, e^{-2}, e^{-3} decays) and then averaged the values for all stations to obtain the mean and standard deviation for each mode. The results, along with values for ${}_0S_2$-${}_0S_5$ (discussed later) are shown in figure 5. We expect the error bars to provide an estimate of the scatter in different measurements of Q^{-1} for each mode. To test this, we plotted the results of Sailor [1978] and Sailor and Dziewonski [1978] as well. They are of two

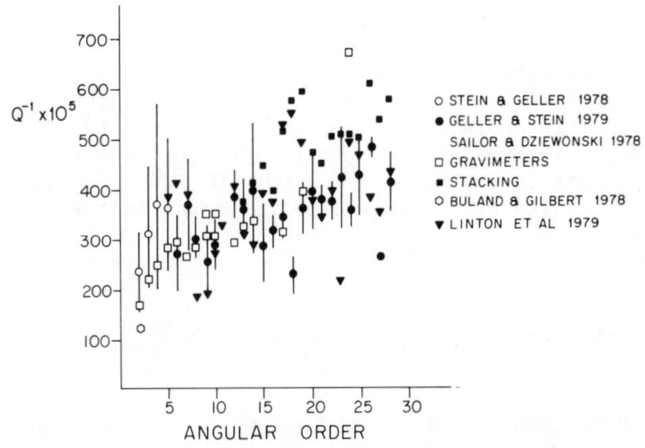

Fig. 5. Fundamental spheroidal mode Q^{-1} data with error bars. Note good agreement with Sailor and Dziewonski's single station gravimeter results (1978), but not the stacking results.

types: single station gravimeter measurements and stacked WWSSN record measurements. It is generally assumed [Sailor and Dziewonski, 1978; Sleep et al., 1981] that stacking tends to overestimate Q^{-1}. This is supported by comparison of stacked results with ours. On the other hand, the single station measurements of Sailor and Dziewonski, and those of Linton et al., [1979], agree with outs. In fact, 75% and 64% of their respective results fall within our error bars, about what would be expected for one-sigma error bars.

The data for the four longest period modes are from Stein and Geller [1978b]. The uncertainities in these data are greater than those for the other modes. These modes are visibly split by rotation and ellipticity into individual peaks [Pekeris, Alterman, and Jarosch, 1961; Backus and Gilbert, 1961]. The resulting time series is a complex beating pattern between different singlets [Stein and Geller, 1978a]. Q^{-1} can no longer be measured directly as for the shorter period modes; it must be estimated from the time series by comparison with synthetic seismograms which include the effects of both splitting and attenuation. As the data for these modes are from the 1960 Chilean and 1964 Alaskan earthquakes, recorded only at single stations on older instruments than those of the IDA network, our error bounds are fairly broad. These error bounds are quite realistic, though, as can be seen by comparison with other studies. The results (tabulated in Stein and Geller, 1978b) of Benioff, Press, and Smith [1961], Alsop, Sutton and Ewing [1961], Slichter [1967], Smith [1972], Sailor and Dziewonski [1978], and Linton, Smylie, and Jensen [1979] all fall within the error bars. Buland and Gilbert's value for ${}_0S_2$ is slightly below them [1978]. The results of a more recent study with IDA data (Stein and Nunn, 1981) in which the attenuation of ${}_0S_3$ and ${}_0S_4$ was measured using both individual singlet decay and multiplet beat patterns are also in excellent agreement.

The full Q^{-1} dataset used for the inversion, with error bars, is shown in figure 6 and tabulated in Table 1. In the period range where the surface wave and normal mode Q^{-1} data overlap the results are averaged. As discussed in the following section, the accuracy with which the data are known is important for the inversion result.

Data Space Inversion

This section discusses the data space inversion method; the following section gives the inversion results. Data space inversion [Backus and Gilbert, 1967, 1968, 1970]is a powerful technique designed to avoid artifacts due to overparameterization of the model space. It is thus especially valuable in inverting for Q^{-1}, for which (unlike velocity inversion) we have only weak a prior knowledge of the structure, because it provides a smoothly varying model with depth rather than introduced layering. We can therefore determine our true resolution with depth.

In the inversion, we used the method of "ranking and winnowing" [Gilbert, 1971] which allows the problem to be formulated using inequality constraints (error bars) on the data. This technique was applied to inversion for Q^{-1} by Sailor [1978] and Sailor and Dziewonski [1978]. Our analysis differs from theirs in that we used only fundamental spheroidal mode data, which we consider the best constrained, over a broader frequency range. As the IDA data were not available at the time of Sailor and Dziewonski's study, our error bounds are probably more representative, as discussed in the previous section. As discussed later, we also used two different criteria to define the quantity to be minimized in perturbing the starting model.

Assuming no bulk dissipation so that Q^{-1} =

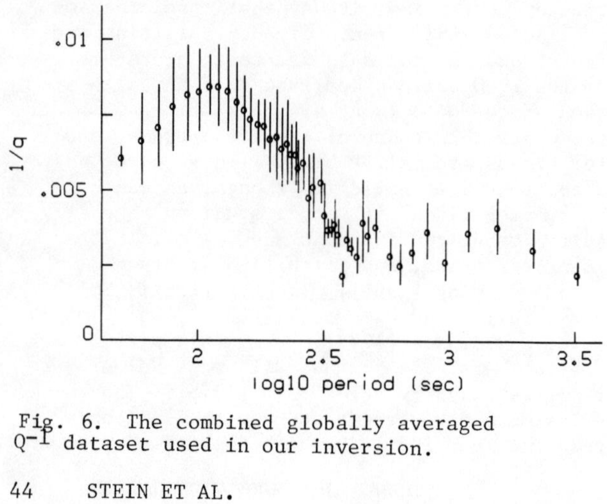

Fig. 6. The combined globally averaged Q^{-1} dataset used in our inversion.

TABLE 1. Q^{-1} and Standard Deviation of Fundamental Spheroidal Modes

Mode	1/Q	σ
OS191	0.00614	0.00034
OS159	0.00671	0.00153
OS136	0.00717	0.00136
OS119	0.00786	0.00154
OS105	0.00826	0.00158
OS 94	0.00835	0.00135
OS 85	0.00853	0.00099
OS 78	0.00851	0.00134
OS 71	0.00837	0.00138
OS 65	0.00801	0.00126
OS 60	0.00776	0.00106
OS 56	0.00744	0.00097
OS 52	0.00728	0.00075
OS 48	0.00722	0.00083
OS 45	0.00679	0.00117
OS 42	0.00687	0.00150
OS 40	0.00646	0.00131
OS 38	0.00664	0.00140
OS 35	0.00628	0.00093
OS 33	0.00626	0.00105
OS 32	0.00583	0.00087
OS 30	0.00599	0.00093
OS 28	0.00483	0.00140
OS 26	0.00520	0.00106
OS 24	0.00533	0.00072
OS 23	0.00425	0.00102
OS 22	0.00377	0.00032
OS 21	0.00378	0.00029
OS 20	0.00395	0.00079
OS 19	0.00362	0.00046
OS 18	0.00224	0.00044
OS 17	0.00345	0.00043
OS 16	0.00316	0.00030
OS 15	0.00287	0.00059
OS 14	0.00401	0.00138
OS 13	0.00358	0.00059
OS 12	0.00385	0.00055
OS 10	0.00290	0.00053
OS 9	0.00258	0.00067
OS 8	0.00303	0.00042
OS 7	0.00370	0.00093
OS 6	0.00272	0.00066
OS 5	0.00367	0.00066
OS 4	0.00385	0.00092
OS 3	0.00313	0.00065
OS 2	0.00230	0.00038

Q_μ^{-1}, we have observations of Q^{-1} for each mode, Q_j^{-1} which are related to the distribution of attenuation with depth by

$$Q_j^{-1} = \int_0^1 G_j(r) Q^{-1}(r) dr \qquad (1)$$

where $G_j(r)$ is the kernel for each mode.

The kernel can be derived from perturbations to the complex mode eigenfrequency [Takeuchi and Saito, 1972], or from the distribution of shear energy with depth [Backus and Gilbert, 1968]

$$G_j(r) = \frac{1}{\omega}\left[\beta(r)\left(\frac{\partial\omega}{\partial\beta}\right)_j(r) + \frac{4\beta^2(r)}{3\alpha(r)}\left(\frac{\partial\omega}{\partial\alpha}\right)_j(r)\right] \quad (2)$$

(This kernel differs by an r^2 normalization from that of Sailor and Dziewonski, 1978).

From a starting model of attenuation with depth, the problem is to minimize the deviation from the starting model

$$\int_0^1 (\delta Q^{-1}(r))^2 \, dr$$

subject to the constraints (1). This is an isoperimetric variational problem, in which an integral is minimized subject to integral constraints (Weinstock, 1974, p. 51). In this case, using Lagrange multipliers, the quantity

$$L = (\delta Q^{-1}(r))^2 + \sum_j \lambda_j G_j(r)\, \delta Q^{-1}(r)$$

must satisfy

$$\frac{\partial L}{\partial(\delta Q^{-1}(r))} = 0 .$$

Thus the solution is obtained by expanding changes in the model in terms of the kernels

$$\delta Q^{-1}(r) = \sum_j \nu_j G_j(r) \quad .$$

This is the "natural" expansion, in that the data are derived along these kernels. The model is a smooth function of depth, containing no arbitrary parametrization. Gilbert [1971] showed how the coefficients ν_j may be chosen to satisfy the inequality constraints

$$\delta Q_j^{-1} - \sigma_j < \int_0^1 G_j(r)\, \delta Q^{-1}(r)\, dr < \delta Q_j^{-1} + \sigma_j \quad (3)$$

where δQ_j^{-1} is the difference between the observed and predicted values of Q_j^{-1} and σ_j is the estimated standard deviation of Q_j^{-1}, by forming the outer product matrix of the kernels

$$A_{ij} = \int_0^1 G_i(r) G_j(r) dr.$$

To weight each mode by its standard deviation, we form

$$H_{ij} = \frac{A_{ij}}{\sigma_i \sigma_j}$$

and the matrix 0_{ij}, whose columns are the eigenvectors of H_{ij}. Finally, letting

$$T_{ij} = \frac{1}{\sqrt{\lambda_i}} \; 0_{ij}^T \; \frac{1}{\sigma_j} \quad (4)$$

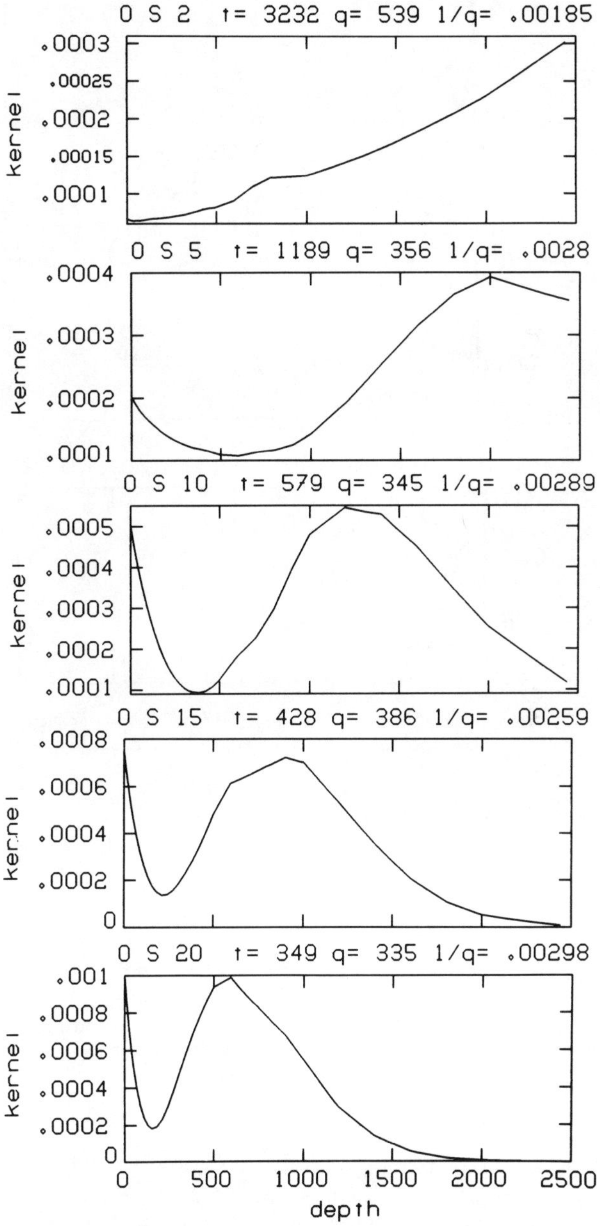

Fig. 7. Q^{-1} kernels for long period modes, derived from earth model 1066A, computed using programs of Wiggins (1976) and our Q^{-1} model.

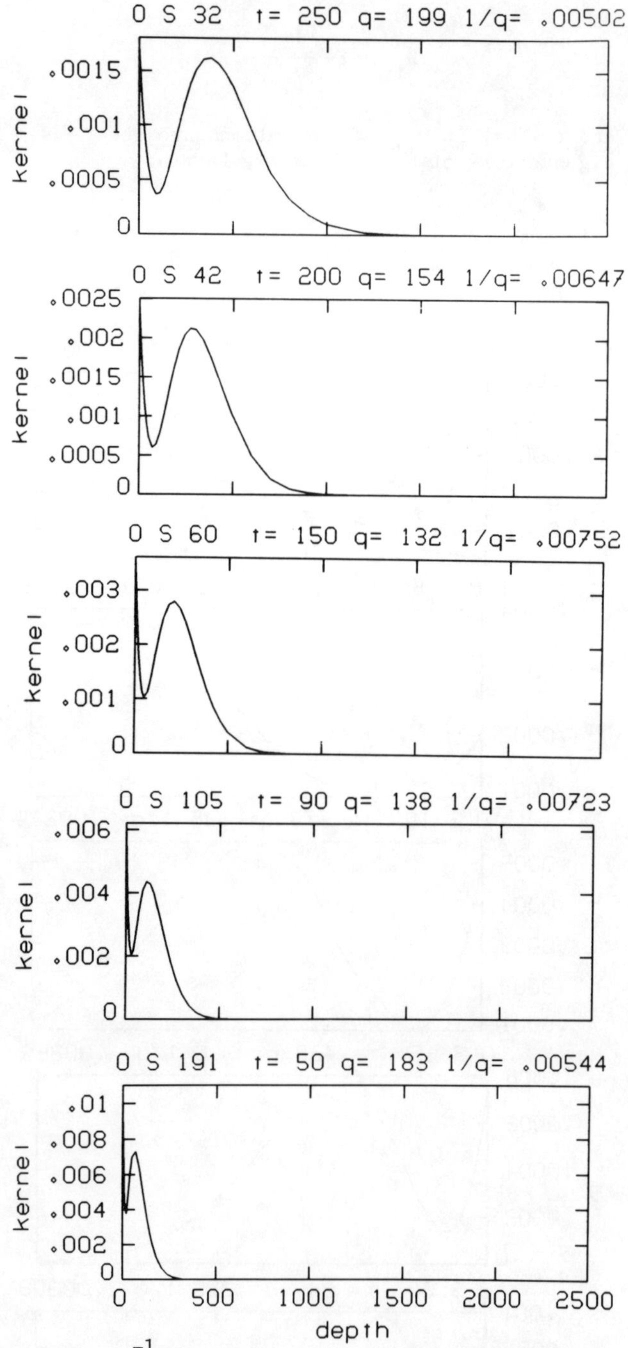

Fig. 8. Q^{-1} kernels for "surface wave" modes, whose energy is concentrated nearer to the surface than the long period modes.

we have a matrix which diagonalizes A_{ij}, $T^T AT = I$. The purpose of this matrix is to expand the model in the transformed kernels $T_{ij}G_j(r)$, (the summation convention is implied)

$$\delta Q^{-1}(r) = \nu_i[T_{ij}\, G_j(r)] \qquad (5)$$

Then, substituting this expression into (3) and multiplying by T_{ij},

$$T_{ij}(\delta Q_j^{-1} - \sigma_j) < \nu_i < T_{ij}(\delta Q_j^{-1} + \sigma_j). \qquad (6)$$

Now, minimizing $(\delta Q^{-1}(r))^2$ dr simply means minimizing $\sum_i \nu_i^2$, so that each ν_i is simply

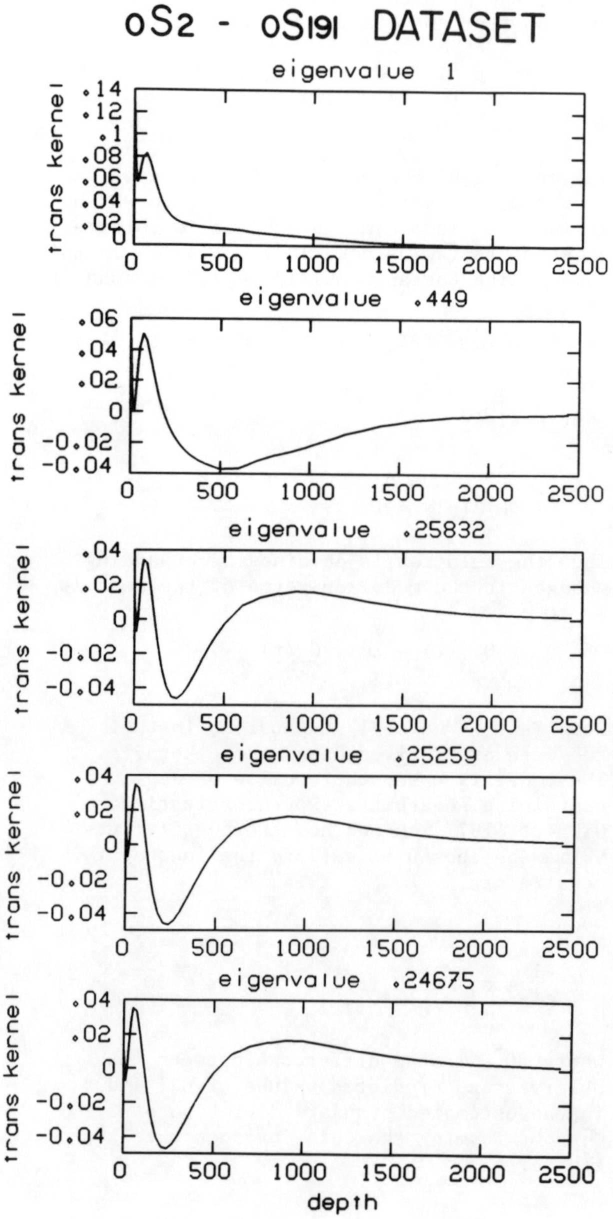

Figure 9. Transformed kernels and associated eigenvalues for the mode dataset. To the (single) precision of the computations, the fourth and fifth kernels are essentially equivalent to the third.

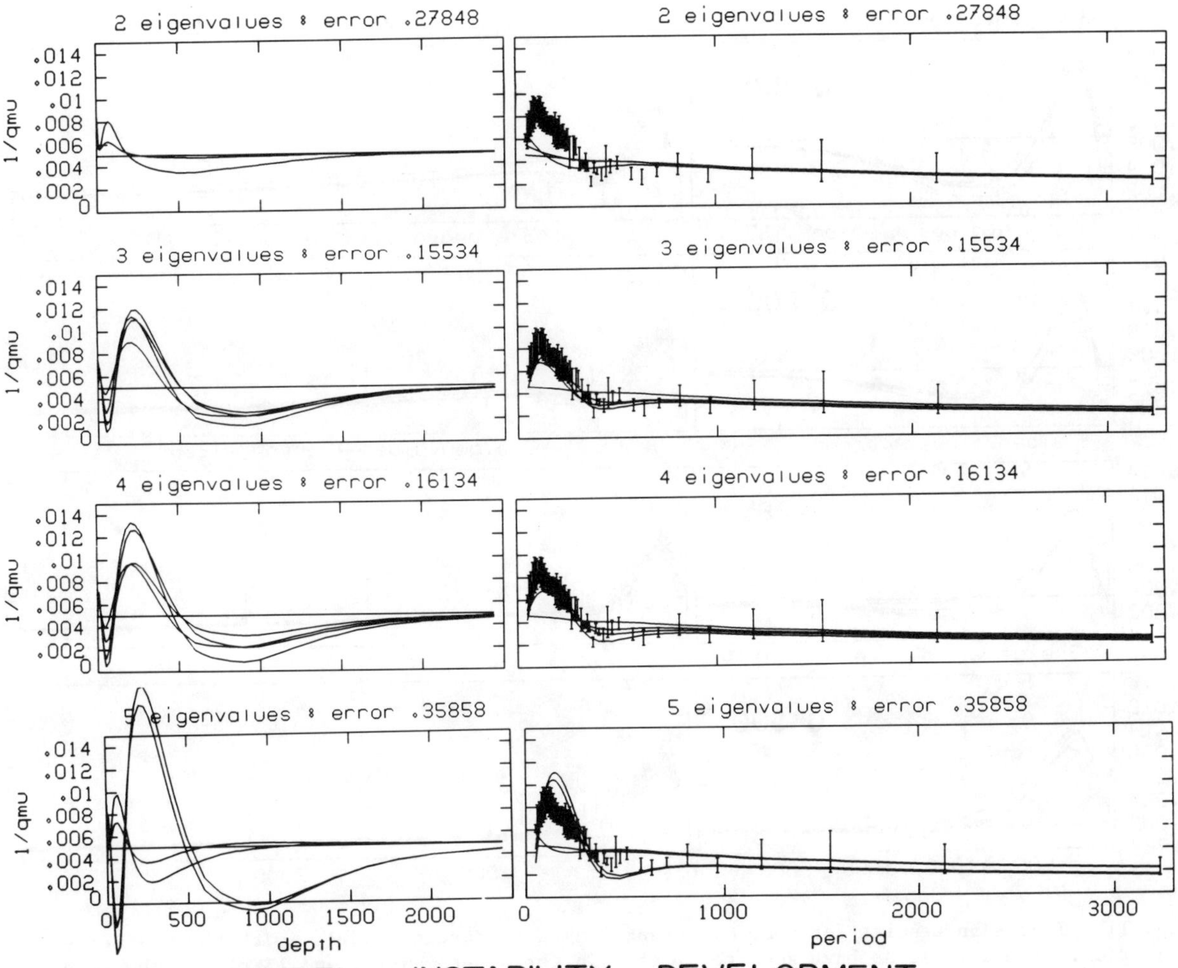

Fig. 10. Inversion, starting with a uniform Q^{-1} model, for different numbers of transformed kernels. The first two do not adequately resolve the attenuation peak associated with the low velocity zone; the first three are quite successful; model instability between successive iterations develops for five kernels. Fits to the data are shown for the starting model (nearly straight line) and for each iteration.

the minimum (absolute value) that satisfies (6).

By expanding the model in terms of the transformed kernels, we achieve several important effects. First, we eliminate problems resulting from the fact that, as shown later, the kernel is a slowly varying function of angular order, which means that many modes have similar kernels. By chosing transformed kernels corresponding to the largest eigenvalues, we use only the most independent combinations of model kernels to form the model. And with a smooth starting model, we ensure that our inversion results will also be smooth. Similarly, by fitting the transformed data, we have scaled all the data to unit standard error and weighted the inversion appropriately.

Inversion Procedure

Figures 7 and 8 show the kernels for some of the modes in our dataset (Table 1) computed using the programs described in Wiggins [1976]. The energy of short period surface waves, down to 50 sec (which corresponds to ${}_0S_{191}$), is concentrated near the surface. Thus, we have excellent resolution of the low velocity (high attenuation) zone. The longer period modes penetrate deeper, and ${}_0S_2$ has large amplitudes down to the core-mantle boundary. As the kernels (equation 2) depend on β, they have no resolution in the outer core. Thus, because we consider only shear dissipation, we restrict our models to the mantle.

The first five transformed kernels for the dataset $T_{ij}G_j(r)$ are shown in Figure 9,

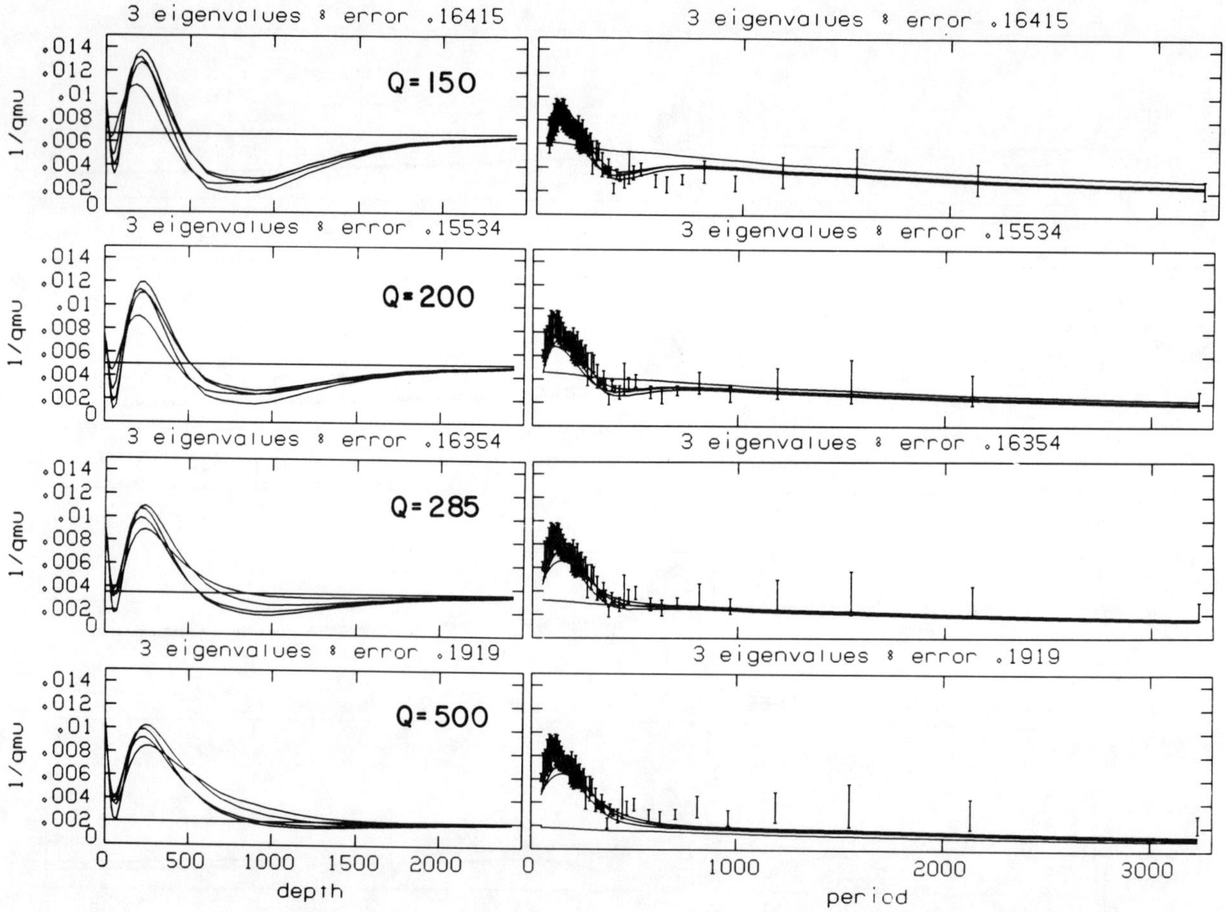

Fig. 11. Inversion results for four different constant starting models, with the uniform scale criterion. The inversions have good resolution in the upper mantle, and little in the lower mantle. The best fit to the data occurs for the Q^{-1} = .005 (Q = 200) starting model.

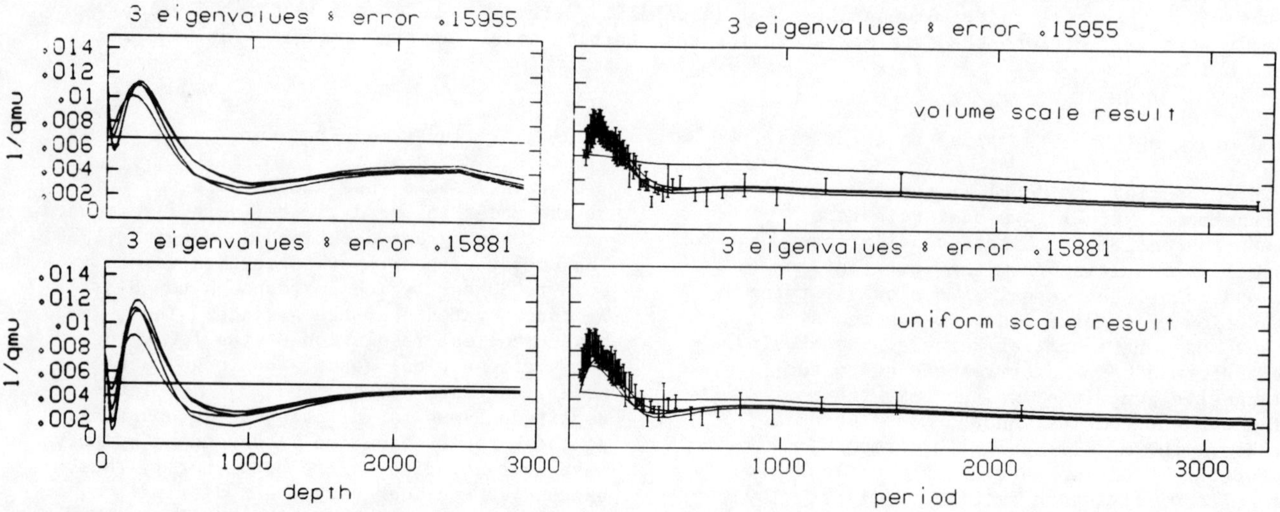

Fig. 12. Comparison of best fit models, derived using the uniform scale and volume scale criteria. The models, though derived from different starting models, are similar and provide equivalent fits to the data. The differences between them indicate a range of equally acceptable models.

with the associated eigenvalues. There are essentially three independent eigenvalues, (and thus transformed kernels). The fourth and fifth kernels are so similar to the third that they do not provide additional resolving power. This can be seen from Figure 10, which contrasts the results of adding successive eigenvalues in the inversion. In each case, the starting model is a uniform $Q^{-1} = .005(Q = 200)$ throughout the mantle. The left panels show the changes in the model; the right panels show the data and the predicted values of Q^{-1}. Q^{-1} from the starting model appears as a roughly straight line fit to the data. Using only the first two eigenvalues, the four successive iterations shown are unable to fit the "bump" associated with the low velocity zone, at periods from 50-200 sec. The three kernel fit, on the other hand, fits the data well. Four kernels yield roughly equivalent results, but successive iterations differ more between themselves. With five eigenvalues the inversion becomes quite unstable between the successive iterations. This instability occurs regardless of starting model. Therefore we have used only three eigenvalues in all of our inversions.

Figure 11 summarizes our inversion procedure. We use constant attenuation starting models, and thus invert for the minimum smooth changes needed to fit the data. By choosing uniform starting models we prevent a priori sharp features from propagating through the inversion, to the final result. The advantages of smooth starting models were pointed out by Sailor [1978] and Sailor and Dziewonski [1978]. In view of the present lack of knowledge of Q^{-1} structure in the earth, we begin with constant Q^{-1} models and perturb them with the inversion.

For a series of four constant Q^{-1} starting models, we derived the results shown in Figure 11. In each case the final models are similar to each other down to about 1200 km, and below that depth become asymptotic to the starting model. Thus, the inversion does not have much resolution at greater depths. In examining Figure 11, it is essential to remember that, unlike parameter space inversion [Wiggins, 1972], different depths in the model cannot be adjusted independently. The inversion proceeds by changing the weight given to each transformed kernel, which requires a change to the entire model. Because of this, and because we are minimizing changes from the starting model, different starting models yield different final models. These different models do not fit the data equally well.

For each final iteration, we show the error criterion,

$$E = \frac{1}{N} \sum_{i=1}^{N} \left(\frac{Q_i^{-1}\ (\text{observed}) - Q_i^{-1}(\text{predicted})}{\sigma_i} \right)^2$$

which shows how well the data are fitted. In all cases the inversion result fits the transformed data (a weighted linear combination), to within the transformed error bounds. The error criterion simply quantifies our observation that inversion results for some starting models fit better than others. In the lower mantle, the inversion has little resolution, since the error bars on the longest period modes are quite broad. Thus a wide range of Q^{-1} models can predict attenuation within one sigma of the data. Examining the results of various starting models, we obtain the best fit for a starting model with $Q^{-1} = .005$ throughout. The various iterations from this starting model form a family of models which yields essentially equivalent fits to the data. We use

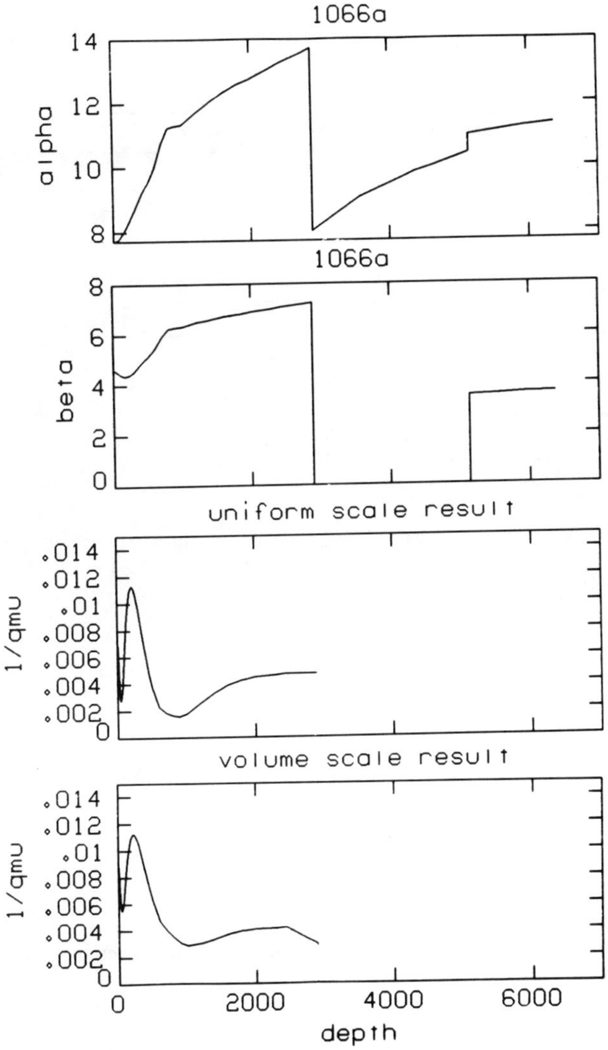

Fig. 13. Best fit models and earth models.

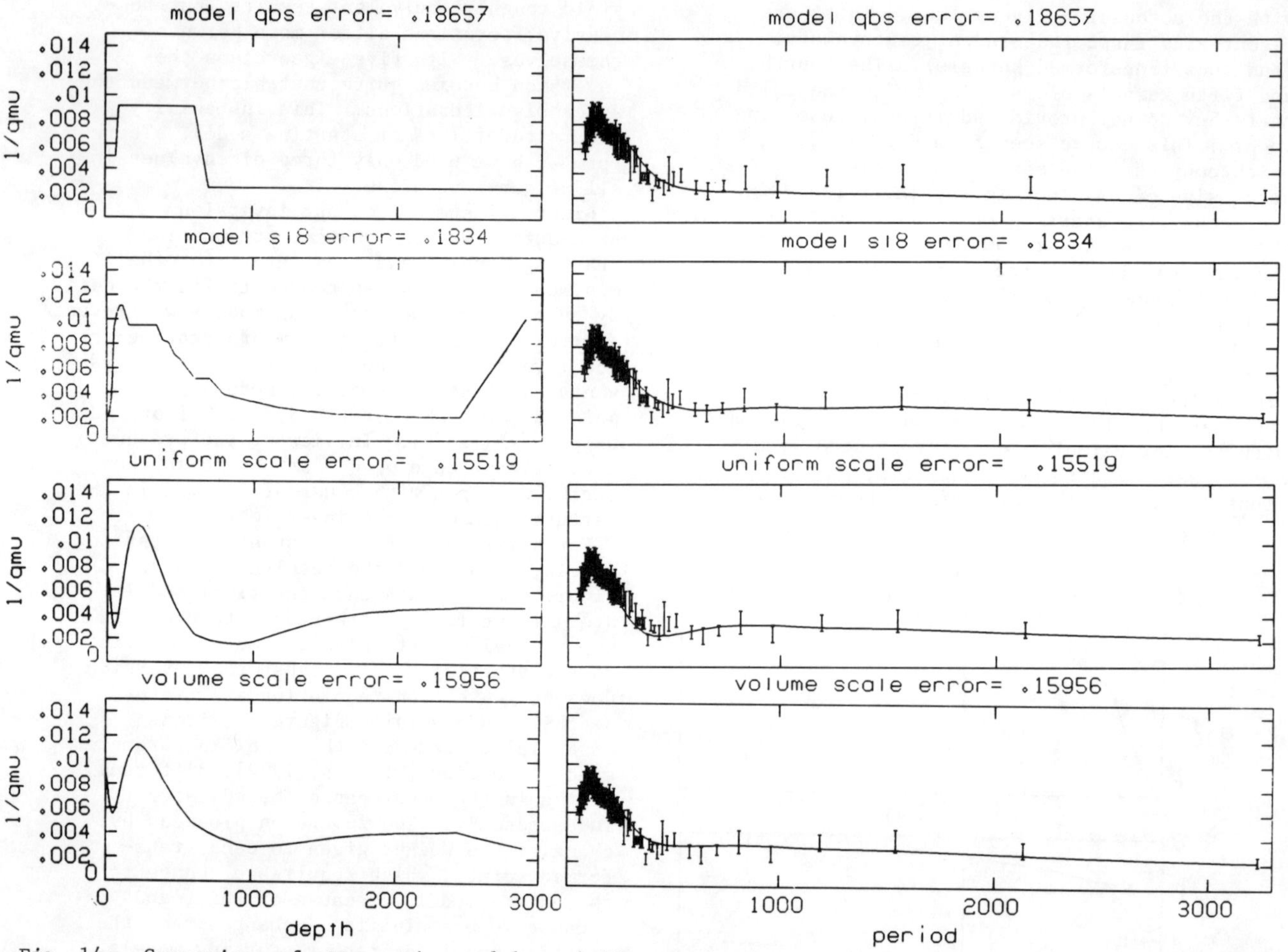

Fig. 14. Comparison of attenuation models and fits to the data. All four models provide acceptable fits and are generally similar. QBS (Sailor and Dziewonski, 1978) and SL8 (Anderson and Hart, 1978) are sampled at the forty depth points used in our inversion.

the final one of this set (Q=200) as representative of our inversion results.

The real limitation on our resolution of Q^{-1} in the lowest mantle is the uncertainty in Q^{-1} for the longest period modes. If we reduce the error bars of ${}_0S_2$ through ${}_0S_5$, the transformed kernels, and thus the inversion, are much more sensitive to the attenuation in the lowest mantle. With what we consider realistic error bounds, our present dataset constrains the upper mantle much more tightly that the lower mantle.

The limited resolution in the lower mantle is primarily a function of having only fundamental mode Q^{-1} data. Similar results were obtained by Nakanshi [1981] who used the same ${}_0S_2$ - ${}_0S_{24}$ data as this study, and his own (Nakanishi 1978, 1979) data for the surface wave band in a parameter space inversion. Even if far more precise fundamental mode Q^{-1} measurements were available, the resolving power of <u>any</u> inversion scheme would not be dramatically better than at present. Ultimately, better models can be obtained only by having good Q^{-1} measurements of spheroidal overtones. Although several overtone Q^{-1} values have been published, they do not seem to be up to the same standards of accuracy as data from IDA, and we have chosen not to use them in the present paper.

Sailor and Dziewonski's [1978] inversions had more resolution than ours in the lowest mantle due, to some extent, to their choice of narrow error bounds for the longest period modes. More importantly, they adopted a different criterion governing deviations from the starting model for the inversion.

Our inversions seek to minimize the changes in the model,

$$\int_0^1 (\delta Q^{-1}(r))^2 \, dr \ ,$$

subject to the constraints imposed by the data. Sailor and Dziewonski [1978] minimized the changes with a weighting function:

$$\int_0^1 (\delta Q^{-1}(r))^2 r^2 dr.$$

This scheme weights the model changes by the associated volumes, since a change at depth involves less material. In a sense, it represents the minimum change in the physical properties of the earth from a given starting model. For this reason, the volume weighting favors changes at depth over changes near the surface. The uniform weighting scheme that we chose does not favor changes at one depth over another, but it lacks the minimum physical properties effect of the other. Naturally, neither scale is "incorrect"; they represent different approaches to the inversion.

We repeated our inversions, using the "volume scale" criterion rather than our original "uniform scale". The outer product matrix, and thus the transformed kernels, differ from the previous ones. As might be expected, in this case the inversion easily constrained the lower mantle, but was somewhat less successful in fitting the attenuation peak (Q minimum) due to the low velocity zone. To achieve a good fit, as measured by the error criterion, we again swept through a suite of starting models and show the one yielding the best result. Figure 12 compares the best fitting uniform scale and volume scale models and shows their fits to the data. The two models are equivalently good fits.

Figure 13, showing the velocity and both attenuation models, demonstrates that the two models are quite similar. In particular, the low velocity zone reaches a maximum attenuation of about .011 (Q = 90) which drops to a broad minimum at about 700-900 km, and rises somewhat in the lower mantle, to about .005 (Q = 200). Above the low velocity zone we have a less attenuating "lid". It is worth noting that the uniform scale result has larger deviations near the surface than the volume scale result, while the volume scale result is less smooth in the lowermost mantle. The uniform scale result has a deeper attenuation minimum at about 700 km and a correspondingly greater attenuation below about 1500 km. Figure 14 compares both models to the three layer QBS model (which also included bulk attenuation) derived by Sailor and Dziewonski [1978]. Our models have somewhat better fits to the data, but the advantage is not major. The primary difference is the simplicity and smoothness of our models, and the fact that they were derived from starting models with completely uniform Q^{-1}, and thus do not include any *a priori* assumptions. Thus the degree to which our models show the low Q zone in the upper mantle LVZ is striking. The numerical values for our two Q^{-1} models in Figure 14 are tabulated in Table 2.

TABLE 2. Q^{-1} Results from Inversion Studies with Two Different Fit Criteria

Radius (km)	Fit Criterion: Uniform	Volume
6371.0	0.007044	0.009015
6361.0	0.006854	0.008235
6351.0	0.005403	0.007135
6341.0	0.003992	0.006236
6331.0	0.003090	0.005699
6321.0	0.002765	0.005520
6301.0	0.003487	0.005971
6281.0	0.005153	0.006985
6261.0	0.007000	0.008131
6241.0	0.008628	0.009179
6221.0	0.009878	0.010027
6191.0	0.010996	0.010867
6161.0	0.011340	0.011233
6131.0	0.011107	0.011217
6101.0	0.010488	0.010920
6071.0	0.009635	0.010429
6031.0	0.008334	0.009602
5991.0	0.007031	0.008706
5951.0	0.005834	0.007828
5911.0	0.004771	0.007006
5871.0	0.003849	0.006248
5771.0	0.002318	0.004772
5671.0	0.001857	0.004029
5571.0	0.001669	0.003511
5471.0	0.001591	0.003061
5371.0	0.001783	0.002857
5171.0	0.002599	0.003038
4971.0	0.003343	0.003374
4771.0	0.003900	0.003712
4571.0	0.004262	0.003954
4371.0	0.004484	0.004098
3931.0	0.004721	0.004213
3484.3	0.004721	0.002959

Conclusion

We regard the two models we show as representative of an entire class of models which are equivalent descriptions of our knowledge of attenuation within the earth. These smoothly varying models, derived by data space inversion starting with the simplest possible models, are adequate to fit fundamental spheroidal mode data over the entire period range 50-3100 seconds.

Acknowledgements

We thank Freeman Gilbert, Richard Sailor and Norm Sleep for helpful discussions, and Hitoshi Kawakatsu for a critical review of the manuscript. This research was supported by National Science Foundation Grants EAR-78-03653 and EAR 80-19463 at Stanford, by National Science Foundation Grant EAR 80-07363 at Northwestern and by the Department of Energy at Livermore.

References

Agnew, D., J. Berger, R. Buland, W. Farrell and F. Gilbert, International deployment of accelerometers: A network for very long period seismology, EOS Trans. Am. Geophys. Un., 57, 180-188, 1976.

Alsop, L.E., G.H. Sutton and M. Ewing, Measurement of Q for very long period free oscillations, J. Geophys. Res. 66, 2911-2915, 1961.

Anderson, D.L. and R.S. Hart, Q of the earth, J. Geophys. Res. 83, 5869-5882, 1978.

Backus, G. and F. Gilbert, Numerical applications of a formalism for geophysical inverse problems, Geophys. J. 13, 247-276, 1967.

Backus, G. and F. Gilbert, The resolving power of gross earth data, Geophys. J. 16, 169-205, 1968.

Backus, G. and F. Gilbert, Uniqueness in the inversion of inaccurate gross earth data, Phil. Trans. R. Soc. Lond. A. 266, 123-192, 1970.

Benioff, H., F. Press and S. Smith, Excitation of the free oscillations of the earth, J. Geophys. Res. 66, 605-619, 1961.

Backus, G. and F. Gilbert, The rotational splitting of the free oscillations of the earth, Proc. Nat. Acad. Sci. US 47, 362-371, 1961.

Buland, R., On interference among free oscillations of the earth, Geophys. J., subm. 1979.

Buland, R. and F. Gilbert, Improved resolution of complex eigenfrequencies in analytically continued seismic spectra, Geophys. J. 52, 457-470, 1978.

Dziewonski, A.M., On regional differences in dispersion of mantle Rayleigh waves, Geophys. J. 22, 289-325, 1970.

Geller, R.J. and S. Stein, Time domain attenuation measurements (${}_0S_6$-${}_0S_{28}$) for the 1977 Indonesian earthquake, Bull. Seism. Soc. Am. 69, 1671-1691, 1979.

Gilbert, F., Ranking and winnowing gross earth data for inversion and resolution, Geophys. J. 23, 125-128.

Gilbert, F. and A.M. Dziewonski, An application of normal mode theory to the retrieval of structural parameters and source mechanisms from seismic spectra, Phil. Trans. R. Soc. Lond. A., 278, 187-269, 1975.

Hamada, K., Regionalized shear-velocities models for the upper mantle inferred from surface wave dispersion data., J. Phys. Earth 20, 301-326, 1972.

Hermann, R.B. and B.J. Mitchell, Statistical analysis and interpretation of surface-wave anelastic attenuation data for the stable interior of North America, Bull. Seism. Soc. Am. 65, 1115-1128.

Jobert, N. and G. Roult, Periods and damping of free oscillations observed in France after sixteen earthquakes, Geophys. J. 45, 155-176, 1976.

Kanamori, H., Velocity and Q of mantle waves, Phys. Earth Planet. Int. 2, 259-275, 1970.

Linton, J.A., D.E. Smylie and O.G. Jensen, Gravity meter observation of free modes excited by the August 19, 1977 Indonesia earthquake, Bull. Seism. Soc. Amer. 69, 1445-1454, 1979.

Luh, P.C., Free oscillations of the laterally inhomogeneous earth: quasi-degenerate multiplet coupling, Geophys. J. 32, 187-202, 1973.

Luh, P.C., Normal modes of a rotating self-gravitating inhomogeneous earth, Geophys. J. 38, 187-224, 1974.

Madariaga, R., Toroidal-free oscillations of the laterally heterogeneous earth, Geophys. J. 27, 81-100, 1971.

Madariaga, R. and K. Aki, Spectral splitting of toroidal-free oscillations due to lateral heterogeneity of the earth's structure, J. Geophys. Res. 77, 4421-4431, 1972.

Mills, J.M. and A.L. Hales, Great circle Rayleigh wave attenuation and group velocity, Part I: Observations for periods between 150 and 600 seconds for 7 great circle paths, Phys. Earth Plan. Int. 14, 109-119, 1977.

Mills, J.M. and A.L. Hales, Great circle Rayleigh wave attenuation and group velocity, Part II: Observations for periods between 50 and 200 seconds for 9 great circle paths and global averages for periods of 50 to 600 seconds, Phys. Earth Plan. Int. 17, 209-231, 1978.

Mitchell, B.J., Surface-wave attenuation and crustal anelasticity in central North America, Bull. Seism. Soc. Am. 63, 1057-1071, 1973.

Mitchell, B.J., L.W.B. Leite, Y.K. Yu and R.B. Herrmann, Attenuation of Love and Rayleigh waves across the Pacific at periods between 15 and 110 seconds, Bull. Seism. Soc. Am. 66, 1189-1201, 1975.

Nakanishi, I., Regional differences in the phase velocity and the quality factor Q of mantle Rayleigh waves, Science 200, 1379-1381, 1978.

Nakanishi, I., Phase velocity and Q of mantle Rayleigh waves, Geophys. J., 58, 35-59, 1979.

Nakanishi, I., Shear velocity and shear attenuation models inverted from the world-wide and pure path average data of mantle Rayleigh waves (${}_0S_{25}$-${}_0S_{80}$) and fundamental spheroidal modes (${}_0S_2$-${}_0S_{24}$), Geophys. J., subm., 1981.

Pekeris, C.L., Z. Alterman, and H. Jarosch, Rotational multiplets in the spectrum of the earth, Phys. Rev. 122, 1692-1700, 1961.

Sailor, R.V., Attenuation of low frequency seismic energy, Ph.D. thesis, Harvard University, Cambridge, Mass. 1978.

Sailor, R.V. and A.M. Dziewonski, Measurements and interpretation of normal mode attenuation, Geophys. J. 53, 559-582, 1978.

Saito, M., Theory for the elastic-gravitational oscillation of a laterally heterogeneous earth, J. Phys. Earth 19, 259-270, 1972.

Sleep, N.H., R.J. Geller and S. Stein, A constraint on the earth's lateral heterogeneity from the scattering of spheroidal mode Q^{-1} measurements, Bull. Seism. Soc. Am. 71, 183-197, 1981.

Slichter, L.B., Spherical oscillations of the earth, Geophys. J. 14, 171-177, 1967.

Smith, S.W., The anelasticity of the mantle, Tectonophysics 13, 601-622, 1972.

Solomon. S.C., Seismic wave attenuation and the state of the upper mantle, Ph.D. Thesis, Mass. Inst. of Technol., Cambridge, Mass., 1971.

Stein, S. and R.J. Geller, Attenuation measurements of split normal modes for the 1960 Chilean and 1964 Alaskan earthquakes, Bull. Seism. Soc. Am. 68, 1595-1611, 1978a.

Stein, S. and R.J. Geller, Time domain observation and synthesis of split spheroidal and torsional free oscillations of the 1960 Chilean earthquake: preliminary results, Bull. Seism. Soc. Am. 68, 325-332, 1978b.

Stein, S. and J. Nunn, Analysis of split normal modes for the 1977 Indonesian earthquake, Bull. Seism. Soc. Am., subm., 1981.

Usami, T., Effect of horizontal heterogeneity of the torsional oscillation of an elastic sphere, J. Phys. Earth 19, 175-180, 1971.

Weinstock, R., Calculus of Variations with Variations with Applications to Physics, Dover, London, 1974.

Wiggins, R.A., The general linear inverse problem: implications of free oscillations and surface waves for earth structure, Rev. Geophys. Space Phys. 10, 251-285, 1972.

Wiggins, R.A., A fast, new computational algorithm for free oscillation and surface waves, Geophys. J. 47, 135-150, 1976.

Q_s OF THE LOWER MANTLE - A BODY WAVE DETERMINATION

I. Selwyn Sacks

Department of Terrestrial Magnetism, Carnegie Institution of Washington, Washington, D. C. 20015

Abstract. Spectral comparison of SKP and S waves from a deep focus earthquake, coupled with a comparison of PKP PKP and P (for a different earthquake) allows the attenuation of the lower mantle S path to be isolated. Q_s for the lower mantle in the frequency range 0.03 to 3 Hz is well constrained to 900 ± 100. This is more than a factor 2 greater than the value from free oscillations (at about 10^{-3} Hz) and indicates a slight frequency dependence of Q.

Introduction

Knowledge of the internal structure of the earth is necessary for an understanding of the tectonic processes which result in phenomena such as plate motion and earthquake activity. One parameter of the internal structure is the anelasticity. Because the anelasticity gives a measure (indirectly) of the relative difficulty of producing motion in the earth, many studies have been made of absorption (1/Q) of seismic waves and free earth oscillations. These are consistent with other seismic data in that we can subdivide the crust and mantle into a high Q lithosphere, a low Q aestheonosphere and a high Q mesosphere. The discussion in this paper is limited to Q_s, the value determined from attenuation of shear waves.

Average values of Q_s for the whole mantle have been determined from spectral ratios of successive reflections of S waves from the core-mantle boundary. The multiple ScS waves are well recorded and the Q values derived in such studies are mainly in the range 160 to 290 (e.g. Yoshida and Tsujiura, 1975; Jordan and Sipkin, 1977; Nakanishi, 1979). These values are consistent with the decay of free earth oscillations, for which a Q_s of about 240 is obtained (Anderson and Hart, 1978b; Sailor and Dziewonski, 1978). Upper mantle Q_s values are lower, but show a strong regional bias. They can be as high as 168 (biased to South America and Japan) but the global average is about 130 (Anderson and Hart, 1978a,b).

For rays which traverse both the upper and lower mantle, the dominant attenuation will be in the upper mantle. In order to calculate lower mantle attenuation with a given accuracy, we would need to know the upper mantle attenuation to a greater accuracy. The difficulty of this situation is increased because the upper mantle Q_s is highly variable laterally due to factors such as lithospheric thickness variations, subduction zones and spreading regions. In addition, the spectral content of the seismic source is unknown and variable from source to source.

The lower mantle Q can be inferred from values of upper mantle and whole mantle Q. Upper mantle (600 km) values range from 110 to 170 with a global average of 130. Combining these with the range reported for the whole mantle, 160-290, gives a possible range of lower mantle values of

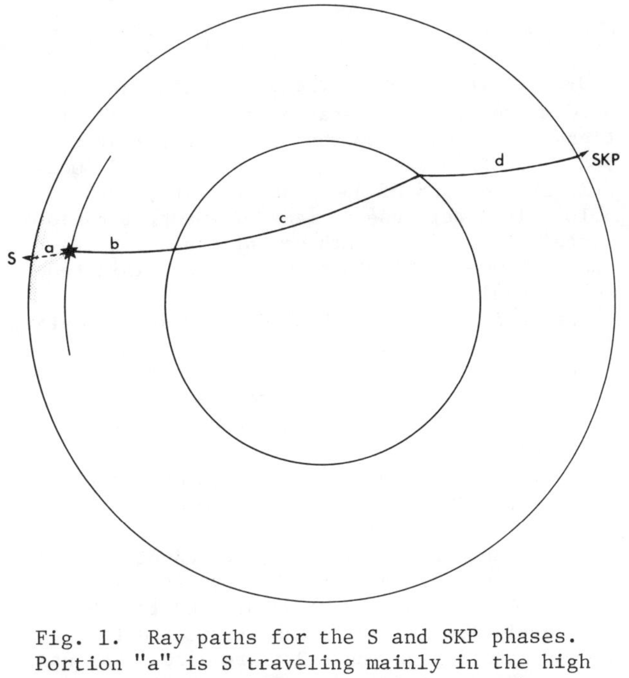

Fig. 1. Ray paths for the S and SKP phases. Portion "a" is S traveling mainly in the high Q subducted lithosphere to a seismograph on the surface. Portion "b" is S traveling from the deep earthquake (star) to the core-mantle interface where it is converted to P. Energy travels through the core, "c", and mantle "d" to the seismograph (SKP) as P waves.

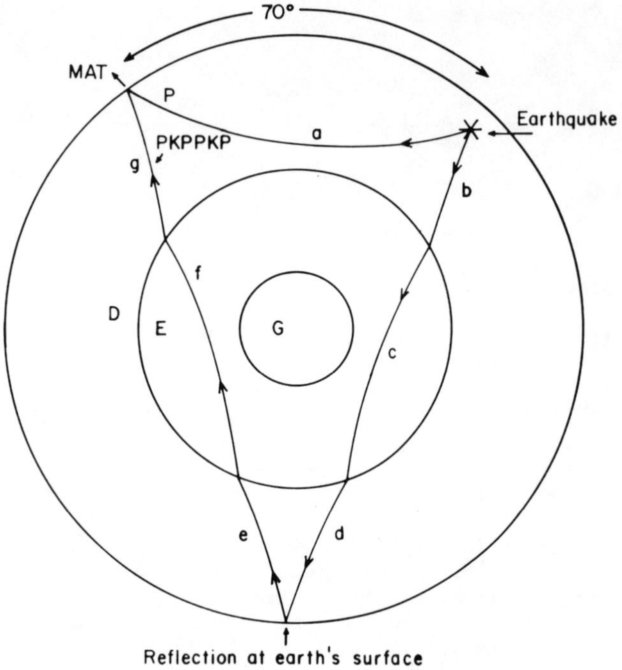

Fig. 2. Ray paths for P and PKP PKP phase. Energy travels from the earthquake to the seismograph through the mantle only, "a", as well as through the core and mantle g, f, e, d, c, b. All paths are P wave paths.

200-750. The lower mantle Q_s can also be determined from the decay of free earth oscillations. Depending somewhat on the particular model, these values range up to about 400 (models SL2, SL7 and SL8 of Anderson and Hart, 1978b).

In this paper, use is made of spectral ratio techniques, and ray path comparisons such that the problems mentioned above can be eliminated or substantially reduced.

The basic philosophy behind the Q determination is as follows:

Spectral ratios of waves traveling two different paths from the same earthquake source are used to cancel out the often unpredictable spectrum of radiation from the source.

In order to minimize the effects of crustal and other structure, which cause oscillations in the recorded spectra, the widest possible bandwidth is analyzed. Frequency dependence of Q is allowed, but not required, by the data presented here, over a frequency range of 0.02 to 3 Hz.

A combination of ray paths is studied to isolate the region where Q is required. By using ScP or SKP waves from deep focus events, the paths through the highly attenuative upper mantle are traversed as P-waves. Compressional waves are less attenuated than shear waves (the ratio of the attenuation of S/P at any frequency is the P attenuation squared, assuming that $Q_p V_p = 3 Q_s V_s$) and therefore their absorption need not be so accurately known.

Method and Data

There are two phases originating from deep earthquakes which have the property of traversing the lower mantle as S but having the final path from the base of the mantle to the seismograph a P wave path. In ScP, S is converted to P at the core-mantle interface and reflected to the surface. It is well observed in the 10-30 degree distance range. While the higher frequency energy (∿3 Hz) is easily determined, frequencies below about 0.3 Hz are obscured by the coda of S.
The SKP arrival (Figure 1) on the other hand, is not contaminated by nearby strong arrivals. This fact, coupled with the equally broad frequency band recording of related waves with pure P paths which allow attenuation on the P section of the path to be determined, made SKP the preferred phase for study.

The spectral ratio of the two arrivals will then be given by

$$\frac{SKP}{S} = \exp \{-\Pi f \, (\frac{Tb}{Qb} + \frac{Tc}{Qc} + \frac{Td}{Qd} - \frac{Ta}{Qa})\} \qquad (1)$$

where a, b, c, d are the segments of the ray paths shown in Figure 1. We are trying to determine Q_b, the lower mantle Q, but Q_a, Q_c and Q_d must be know first.

Q_a has been determined independently for the seismograph-earthquake region used here (Sacks and Okada, 1974). Its ray path is almost

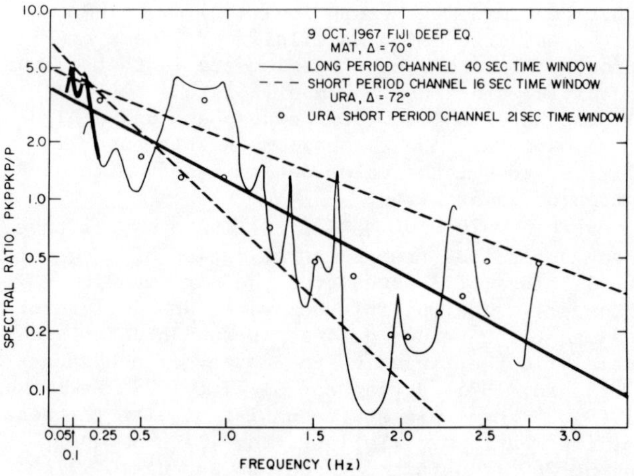

Fig. 3. Spectral ratio of PKP PKP/P determined on two different seismographs. MAT is in central Honshu; KMU is in Hokkaido, Japan. Circles are data from KMU. The dashed lines were used to provide values for the limiting cases which illustrate the sensitivity of the final result to this measurement (see text). The full line is the preferred fit to the data.

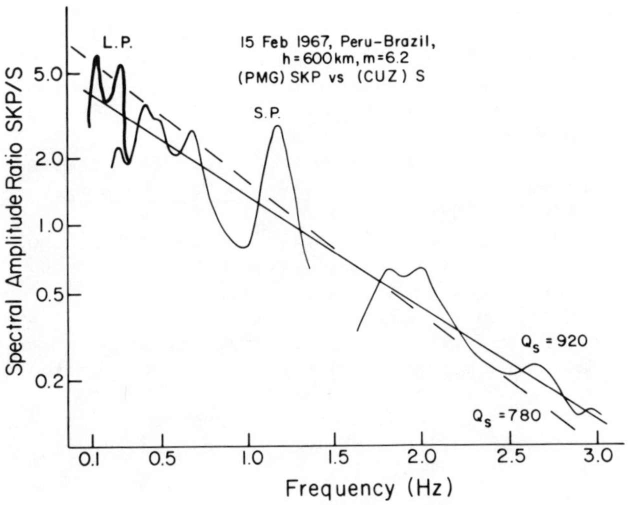

Fig. 4. Spectral ratio SKP/S used to determine the Q_s of the lower mantle (below 600 km). The ratio is plotted only where the signal to noise ratio of both phases is greater than 3:1. The noise is determined using a window identical to that used for the data on a section of record immediately preceding it. The broken line overestimates the low frequency amplitude and underestimates the high frequencies. Yet the resultant Q value only changes 15%. This illustrates the stability of the determination.

entirely within a high-Q subduction plate with $Q_s \simeq 500$. Q_c and Q_d, the outer core and whole mantle P wave paths are determined by studying the ray paths shown in Figure 2.

$$\frac{\text{PKP PKP}}{\text{P}} = \exp\left\{-\Pi f\left(\frac{Tb}{Qb} + \frac{Tc}{Zc} + \frac{Td}{Qd} + \frac{Te}{Qe} + \frac{Tf}{Qf} + \frac{Tg}{Qg} - \frac{Ta}{Qa}\right)\right\}$$

$$= \exp\left\{-\Pi f\left(\frac{Tb}{Qb} + \frac{Tg}{Qg} - \frac{Ta}{Qa} + 2\left(\frac{Tc}{Qc} + \frac{Td}{Qd}\right)\right)\right\} \tag{2}$$

We now show that the term $\left(\frac{Tc}{Qc} + \frac{Td}{Qd}\right)$ dominates. Ta - (Tb + Tg) is about 90 seconds whereas 2 (Tc + Td) is about 1,750 seconds. "a, b, d, g" are whole mantle paths with slightly different geometries and will have similar Q_p values. The core Q_p is rather higher. Substituting reasonable values for these parameters (Q_p = 2000 for lower mantle, 10,000 for outer core paths (Sacks, 1971)) we calculate that

$$\left(\frac{Tb}{Qb} + \frac{Tg}{Qg}\right) - \frac{Ta}{Qa} \approx 0.045 \text{ s}$$

and

$$2\left(\frac{Tc}{Qc} + \frac{Td}{Qd}\right) \approx 0.4 \text{ s}$$

Therefore equation (2) may be reduced to

$$\frac{\text{PKP PKP}}{\text{P}} = \exp\left\{-\Pi f 2\left(\frac{Tc}{Qc} + \frac{Td}{Qd}\right)\right\} \tag{3}$$

with sufficient accuracy. Referring to equation (1), $\left(\frac{Tc}{Qc} + \frac{Td}{Qd}\right)$ is what was required to allow the determinations of lower mantle Q_s without further assumptions.

All the data used was recorded on DTM broad band wide dynamic range magnetic tape recording seismographs (Sacks, 1966). The instrument range extends below 50 second period and above 10 Hz.

For the S/SKP spectral comparison, the S phase of a deep earthquake (600 km) in the Peru-Brazil region of South America was recorded on the CUZ seismograph, and the SKP phase was recorded on the PMG, New Guinea, seismograph. For the PKP PKP/P comparison the earthquake was a large, deep event in the Fiji region, and both arrivals were recorded at the MAT seismograph in central Honshu, Japan, as well as at KMU in Hokkaido, Japan. These data were selected because they had a particularly good signal-noise ratio over a wide frequency range.

Results

The spectral ratio PKP PKP/P, which determines the absorption of the P wave paths, is shown in Figure 3. The frequency range over which the arrivals were above noise was 0.02 to 3 Hz. Figure 4 shows the SKP/S spectral ratio determined over a similar range.

We now determine the permitted range of lower mantle Q_s. The Q_s value (920) determined for the lower mantle is not particularly sensitive to any assumption made but does depend on attenuated estimates of three paths shown in Figure 1. For the subducted lithosphere portion of the path, ("a" in Figure 1), Q_s values of 750 and 350 rather than the 500 actually determined (Sacks and Okada, 1974) are consistent with lower mantle Q_s values of 800 and 1150.

For the P portion of the SKP path, the envelope of extreme values is shown in Figure 3. Minimum P wave attenuation (minimum slope of spectral ratio) is compatible with Q_s for the lower mantle of 760. Maximum P wave attenuation (maximum slope) yields Q_s of 1140.

The fundamental ratio of course is SKP/S. It can be seen in Figure 4 that a change in Q of 15% is precluded by the data. For a Q_s of 780, for example, all the longer period amplitudes (f < 1 Hz) are too low and the shorter period amplitudes are above the calculated values.

The values given above are the extremes, the probable range of Q_s of the lower mantle (below 600 km) is 900 ± 100.

Discussion

The Q_s for the lower mantle determined from the decay of free earth oscillations (at a frequency ∿ .001 Hz) is no more than 400, which is significantly different from the value determined in this study. This implies a slight but definite dependence of Q in the lower mantle.

References

Anderson, D. L., and R. S. Hart, Attenuation models of the earth, Phys. Earth Planet. Interiors, 16, 289-306, 1978a.

Anderson, D. L., and R. S. Hart, The Q of the earth, J. Geophys. Res., 83, 5869-5882, 1978b.

Jordan, T. H., and S. A. Sipkin, Estimation of the attenuation operator for multiple ScS waves, Geophys. Res. Letters, 4, 167-170, 177, 1977.

Nakanishi, I., Attenuation of multiple ScS waves beneath the Japanese arc, Phys. Earth Planet. Interiors, 19, 337-357, 1979.

Sacks, I. S., A broad-band large dynamic range seismograph, Geophys. Monogr., 10, Am. Geophys. Union, 543-553, 1966.

Sacks, I. S., Core phase nomenclature: suggested changes, Carnegie Inst. Wash. Year Book 69, 413-414, 1971.

Sacks, I. S., and H. Okada, A comparison of the anelasticity structure beneath western South America and Japan, Phys. Earth Planet. Interiors, 9, 211-219, 1974.

Sailor, R. V., and A. M. Dziewonski, Measurements and interpretation of normal mode attenuation, Geophys. J., 53, 559-581, 1978.

Yoshida, M., and M. Tsujiura, Spectrum and attenuation of multiple reflected core phases, J. Phys. Earth, 23, 31-42, 1975.

THE VISCOSITIES OF THE EARTH'S MANTLE

W. R. Peltier, P. Wu, and D. A. Yuen

Dept. of Physics, University of Toronto, Toronto, Ontario, Canada M5S 1A7

Abstract. Tensor constitutive relations are deduced for several linear visco-elastic models of the mantle and it is suggested that, of these, the generalized Burger's body warrants most serious consideration as a complete model of mantle anelasticity. Different phenomena "see" different parameters of the model and the spectrum of observed anelastic effects is sufficiently broad that all parameters should be well constrained by geophysical data. The phenomenon of postglacial rebound, in particular, is sensitive only to the long timescale viscosity of the planet. We present analyses here which strongly suggest that this viscosity is independent of time and therefore that transient rheology plays no role in glacial isostasy. When free air gravity data are combined with relative sea level information, the viscosity of the lower and upper mantles are both well constrained. The viscosity contrast between these two regions is small and the value of the mean viscosity is that required by the thermal convection hypothesis of continental drift.

Introduction

The problem of the rheology of the planetary interior continues to be a rich source of geophysical controversy. Such controversy has perhaps been made more acute by the recent recognition by seismologists (eg. Liu et. al., 1976) of the validity of Jeffrey's longstanding claim that the anelasticity of the Earth should not be neglected in elastic wave propagation problems. Direct evidence for anelastic behaviour of the Earth's mantle and crust is abundant, both for phenomena with short characteristic timescale (dispersion of body wave velocities, spatial attenuation of surface waves, finite Q of the elastic gravitational free oscillations), and for longer timescale processes, (post-glacial rebound, crustal bending, polar wander). An equally important piece of a priori evidence for steady state creep processes in the mantle is that the region is mixed by convective overturning. This process could not occur unless it were possible for mantle material to deform continuously in a fluid like fashion in response to an applied shear stress.

Given that anelasticity is important, how are we to describe it? The answer to this question may depend upon the use to which the description is to be put. For certain purposes, such as the interpretation of deformation textures, a complete microphysical description of the process(es) through which the material deforms under an applied stress may be necessary. For other purposes, a macroscopic description in terms of a constitutive relation between stress and strain may be sufficient. In order to construct mathematical models for the anelastic effects mentioned in the first paragraph a macroscopic description will clearly suffice. Unfortunately, this does not simplify the theoretical problem since different microscopic processes (in general) imply different macroscopic constitutive relations. Since it is very difficult in the laboratory to accurately reproduce the material and thermodynamic environment of the mantle, and so to measure its rheological properties directly, we are more or less obliged to proceed by model fitting. That is, given a plausible rheological model we use it to predict the known set of anelastic observables and thus to test its applicability. We will attach special significance, if only from a utilitarian point of view, to the simplest macroscopic constitutive relation which will allow us to understand the complete spectrum of anelastic processes observed. For present purposes we shall assume that the simplest model is the one with the fewest adjustable parameters. Once we have found a sufficiently simple model that it provides an economical means of summarizing anelastic effects we might hope (!) to find microphysical justification for it. This process should be viewed as complementary to the process which begins with a microphysical description of the deformation mechanism. For the most part the present paper will be confined in scope to a discussion of macroscopic descriptions of mantle anelasticity.

In the past few years there has apparently been considerable convergence of view as to what might constitute a sensible working model of viscoelastic behaviour, although this has yet to lead to any consensus concerning a rheological model for the Earth. Such agreement as does currently exist has been marked by the increasing recognition of the utility of linear viscoelastic models in a wide range of geophysical applications. Such models have been most fully exploited in the context of analyses of the postglacial rebound problem (Peltier 1974, 1976; Peltier and Andrews, 1976; Peltier et al., 1978; Wu and Peltier, 1981 a, b) where Correspondence Principle methods have been shown to reduce the complexity of calculation enormously. All of this work has been based upon the use of a linear steady-state Maxwell rheology and the validity of the model therefore hinges upon the assumption that the relaxation processes involved in isostatic adjustment are governed by the same viscosity as controls the longer timescale mantle convection process. The viscosity which one obtains for the upper mantle by using this model to invert the rebound data is on the order of 10^{22} Poise (c.g.s. units) and this viscosity is just that which is required in the context of the convection hypothesis of continental drift (Peltier 1980 a,b). Insofar as the compatibility of this macroscopic model with known microphysical processes is concerned, the situation is still in somewhat of a state of flux. Although much laboratory data favours a non-Newtonian creep law (Kohlstedt and Goetze, 1974; Kohlstedt et al, 1976) there is mounting evidence that for low stress levels such as are involved in mantle convection and postglacial rebound ($\lesssim 10^2$ bars) the creep mechanism may very well be Newtonian if the deformation mechanism involves structural superplasticity (Twiss, 1976; Berckhemer, 1979). Most of this paper will be concerned with further discussion of the linear visco-elastic Maxwell model. In Section 3 we will review recent attempts to demonstrate that transient rheologies are very difficult to reconcile with postglacial rebound data. Section 4 is devoted to a discussion of the way in which the resolution of lower mantle viscosity by isostatic rebound data is enhanced by the use of free air gravity as well as relative sea level information.

Another area in which linear visco-elastic models have enjoyed a steadily increasing recent popularity concerns the study of seismic "Q". The temporal decay of the free oscillations, the spatial decay of propagating surface waves, and the observed dispersion of body wave velocities are all effects due to short timescale anelasticity. Given the success of the linear Maxwell model in reconciling long timescale viscoelastic phenomena, one might be tempted to employ it in the seismic regime also. When this is done, however, it is necessary to use a numerical value for the viscosity which is radically different from the one found by fitting the model to postglacial rebound data. Only then can the model be forced to deliver reasonable values for the Q's of the normal modes (although it cannot give the observed $Q(\omega)$). This situation is clearly unsatisfactory--the model which fits rebound data cannot simultaneously fit the decay of the free oscillations. Yet we are reluctant to allow that the viscosity of the Maxwell model may not be a valid physical property of the Earth, a view in which we are encouraged by virtue of the fact that the Maxwell viscosity is just that needed in the convection hypothesis. It is therefore interesting to note that recently constructed linear visco-elastic models for the study of seismic Q (Liu et al., 1976; Anderson et al., 1976; Anderson and Hart, 1978; Anderson and Minster, 1979) have been based upon the constitutive relation for a standard linear solid rather than for a Maxwell solid. This model is of course natural to the seismic problem since it is characterized by transient anelasticity rather than by the steady state creep of the Maxwell model. Anderson and Minster (1979) have recently suggested a particular form of the standard linear solid in which the effective viscosity is time dependent. They argue that the dominant mechanism of attenuation in a solid whose temperature is in excess of one-half the melting temperature is the so-called high temperature background (HTB), the possible importance of which in the mantle had been suggested by Anderson (1967) and Jackson and Anderson (1970). This mechanism leads to a linear stress strain relation in which the effective viscosity has a power law dependence upon time. There exists considerable laboratory support for such a transient creep law (Goetze, 1971; Goetze and Brace, 1972; Murrell and Chakravarty, 1973; Berckhemer et al., 1979) and also some theoretical justification (Mott, 1953). In their discussion of this model, however, Anderson and Minster (1979) have suggested that the timescale of the transient behaviour is extremely long, indeed so long that the phenomenon of postglacial rebound would lie within it. As we shall show in Section 4, the data from postglacial rebound provide no support for this notion and in fact reject it strongly. Insofar as the problem of constructing an acceptable rheological model is concerned, we are therefore in a predicament. The Maxwell model which works for long timescale phenomena does not work for seismic processes; similarly, the standard linear solid which one can use to fit data on seismic Q has a steady state viscosity which is infinite and it is therefore inappropriate for long timescale processes.

Although the particular form of the standard linear solid proposed by Anderson and Minster is demonstrably inappropriate over the timescales on which they claim it to be valid (Sect. 4), the notion that a standard linear analogue (with a distribution of relaxation times associated with it) might provide the most appropriate vehicle for summarizing short timescale anelasticity is a very useful one. When coupled with the fact that long timescale viscoelasticity is very well described by the steady state Maxwell model we see immediately

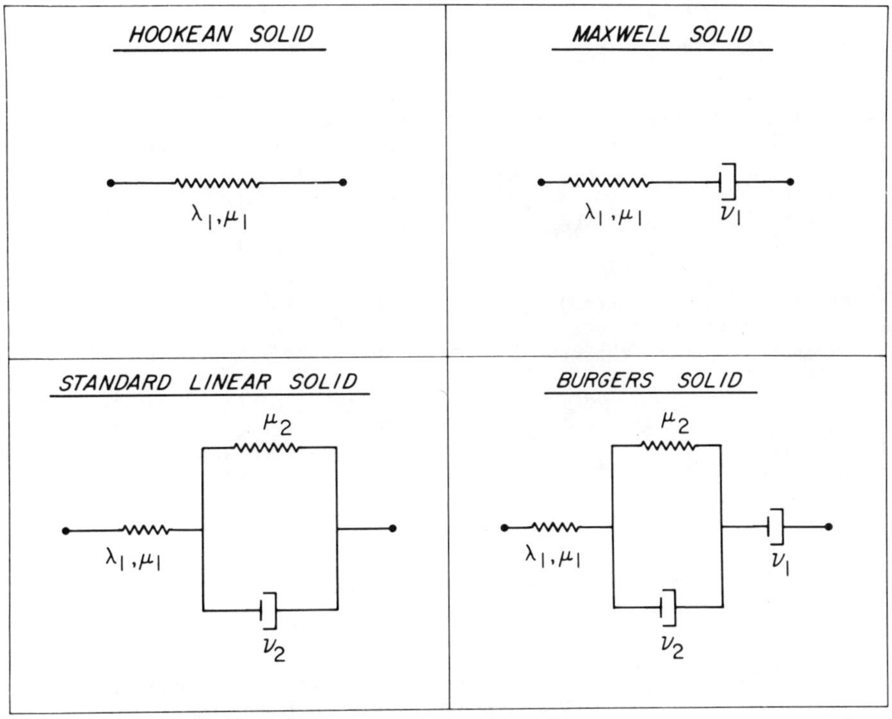

Fig. 1 One dimensional spring and dashpot analogues for the four simple linear viscoelastic solids discussed in the text.

that the simplest linear visco-elastic model in terms of which we might hope to describe the full range of observed anelastic effects is the so-called Burger's body (e.g. Gittus, 1975). This model, which in its simplest form consists of a superposition of standard linear and Maxwell elements, is characterized by two shear viscosities, one which governs short timescale processes, and one which controls the rate of steady state deformation. The next section provides a cursory discussion of linear visco-elastic earth models in general and of the Burger's body in particular. This model deserves to be thoroughly tested against existing observations of anelastic behaviour to determine whether or not it might provide a self-consistent description. Preliminary evidence suggests that it will prove to be a very useful model indeed.

Linear visco-elastic constitutive relations

In Figure 1 are shown the standard one-dimensional spring and dashpot analogues of the linear visco-elastic models which we shall be concerned with here. Three-dimensional tensor forms for the constitutive relations of these rheological models may be deduced using well known methods which are described, for example, in Malvern (1969). For the simple Burger's body the constitutive relation between the stress tensor $\tau_{k\ell}$ and the strain tensor $e_{k\ell}$ is

$$\ddot{\tau}_{k\ell} + \frac{(\mu_1 + \mu_2)}{\nu_2} + \frac{\mu_1}{\nu_1}\left(\dot{\tau}_{k\ell} - \frac{1}{3}\dot{\tau}_{kk}\delta_{k\ell}\right) + \frac{\mu_1\mu_2}{\nu_1\nu_2}\left(\tau_{k\ell} - \frac{1}{3}\tau_{kk}\delta_{k\ell}\right) \tag{1}$$

$$= 2\mu_1\ddot{e}_{k\ell} + \lambda_1\ddot{e}_{kk}\delta_{k\ell} + \frac{2\mu_1\mu_2}{\nu_2}\left(\dot{e}_{k\ell} - \frac{1}{3}\dot{e}_{kk}\delta_{k\ell}\right)$$

where the dot denotes time differentiation. In (1) μ_1 and λ_1 are the instantaneous Lamé parameters. The two viscosities ν_1 and ν_2 are respectively the long timescale and short timescale parameters and for this simple model it has clearly been assumed that there is only a single relaxation time associated with the Kelvin-Voigt element. From (1) we may deduce the constitutive relations for the simpler Maxwell and standard linear models. For example, if we take the limit $\nu_2 \to \infty$ in (1) we obtain

$$\ddot{\tau}_{k\ell} + \frac{\mu_1}{\nu_1}\left(\dot{\tau}_{k\ell} - \frac{1}{3}\dot{\tau}_{kk}\delta_{k\ell}\right)$$

$$= 2\mu_1\ddot{e}_{k\ell} + \lambda_1\ddot{e}_{kk}\delta_{k\ell}$$

which can be integrated once in time to give the Maxwell constitutive relation

$$\dot{\tau}_{k\ell} + \frac{\mu_1}{\nu_2}\left(\tau_{k\ell} - \frac{1}{3}\tau_{kk}\delta_{k\ell}\right)$$

$$= 2\mu_1 \dot{e}_{k\ell} + \lambda_1 \dot{e}_{kk}\delta_{k\ell} \qquad (2)$$

which was introduced by Peltier (1974) in developing the visco-elastic model of glacial isostatic adjustment. If in (1) we take the opposite limit $\nu_1 \to \infty$ then we obtain the new constitutive relation

$$\ddot{\tau}_{k\ell} + \frac{(\mu_1 + \mu_2)}{\nu_2}\left(\dot{\tau}_{k\ell} - \frac{1}{3}\dot{\tau}_{kk}\delta_{k\ell}\right)$$

$$= 2\mu_1 e_{k\ell} + \lambda_1 e_{kk}\delta_{k\ell}$$

$$+ \frac{2\mu_1\mu_2}{\nu_2}\left(\dot{e}_{k\ell} - \frac{1}{3}\dot{e}_{kk}\delta_{k\ell}\right)$$

which may also be integrated once in time to yield the constitutive relation for the simple standard linear solid as

$$\dot{\tau}_{k\ell} + \frac{(\mu_1 + \mu_2)}{\nu_2}\left(\tau_{k\ell} - \frac{1}{3}\tau_{kk}\delta_{k\ell}\right)$$

$$= 2\mu_1 \dot{e}_{k\ell} + \lambda_1 \dot{e}_{kk}\delta_{k\ell}$$

$$+ \frac{2\mu_1\mu_2}{\nu_2}\left(e_{k\ell} - \frac{1}{3}e_{kk}\delta_{k\ell}\right) \qquad (3)$$

If in (2) we take the limit $\nu_1 \to \infty$ or in (3) the limit $\nu_2 \to \infty$ then in either case we obtain, after time integration, the following constitutive relation for a Hooke's law solid

$$\tau_{k\ell} = 2\mu_1 e_{k\ell} + \lambda_1 e_{kk}\delta_{k\ell} \qquad (4)$$

Equations (1) - (4) are the stress-strain relations for the three-dimensional versions of the spring and dashpot analogues shown in Figure 1. These models are certainly not the most general linear visco-elastic models conceivable. We might, for example, wish to allow for the effect of bulk dissipation which has been neglected in constructing (1) - (3).

The Burger's body described by (1) is specified by five parameters $(\lambda_1, \mu_1, \nu_1, \mu_2, \nu_2)$ rather than the two parameters (λ_1, μ_1) which are required to describe the Hookean elastic solid (4). As pointed out above, however, two of these parameters are in common with the Hookean solid (λ_1, μ_1) and these are now well known functions of radius determined (with density) by the systematic inversion of free oscillation data. Furthermore the steady state viscosity ν_1 is known from studies of postglacial rebound. The remaining variables (μ_2, ν_2) are clearly to be determined by fitting the model to short timescale anelastic processes including the Q's of the elastic gravitational normal modes. Results from a sequence of comparisons of the Burger's body model with various anelastic phenomena will be described elsewhere. In the derivation of the above constitutive relations we have nowhere assumed that the viscosities ν_1, ν_2 are functions of the spatial co-ordinates only. They could also be time dependent in which case the rheologies which they describe might properly be called transient. When such time dependence is allowed it is intended as a linear parameterization of an inherently non-linear rheology in which the strain rate is not a linear function of the stress. The extent to which such rheologies might be allowed by the rebound data will be discussed in the next section.

Before beginning this discussion of transient rheology, however, there is one important question which arises concerning the applicability of the general Burger's body model to postglacial rebound. Let us assume that the short timescale viscosity of the model is actually fixed by fitting it to the Q's of the elastic gravitational normal modes of free oscillation. One cannot in fact fit the single Debye peak model to the totality of these data since the model predicts Q's which increase strongly with frequency. This may be corrected by describing the short time scale anelasticity in terms of an absorption band model which includes a spectrum of relaxation times (e.g. Gross, 1953; MacDonald, 1961; Liu et al., 1976) since this model yields values of Q which are not strongly frequency dependent (Yuen and Peltier, 1981). In the Laplace transform domain the constitutive relation for a general absorption band model may be written as (e.g. Christensen, 1971)

$$\tilde{\tau}_{ij}(s) = \delta_{ij}\phi(s)s\,\tilde{e}_{kk}(s) + 2\Psi(s)s\,\tilde{e}_{ij}(s) \qquad (5a)$$

where $\phi(s)$ and $\Psi(s)$ are appropriate relaxation functions (Laplace transforms of the relaxation time distributions, e.g. (Gross, 1953)). For a generalized Burger's body with a continuous anelastic relaxation spectrum which has parabolic shape but which has no bulk dissipation the functions $\Psi(s)$ and $\phi(s)$ are

$$\Psi(s) = \frac{\mu_1}{s + \mu_1/\nu_1}\left(1 + \frac{2}{\pi Q_m}\ln\left(\frac{s + 1/T_2}{s + 1/T_1}\right)\right)$$

$$\phi(s) = \frac{K}{s} - \frac{2}{3}\Psi(s)$$

where K is the unrelaxed bulk modulus. Although a model like (5a) is required to fit the $Q(\omega)$ spectrum we may still employ the single Debye peak model to estimate an upper bound on the short term viscosity by fitting to the lowest frequency (i.e. fundamental) spheroidal and toroidal modes. When this is done (Yuen and Peltier, 1981) one finds that values of ν_2 which are on the order of 10^{17} Poise are required. Since this viscosity is orders of magnitude smaller than that which controls postglacial rebound it is clear that insofar as phenomena on seismic timescales are concerned, the Earth model will behave as a standard linear solid as assumed by Anderson and Minster (1979) and others. With $\nu_2 << \nu_1$ it is very much less clear, however, that the Burger's body model will behave like a Maxwell solid for long timescale processes like postglacial rebound. It might be naively expected that the relaxation spectrum (Peltier, 1976) might be strongly effected by the existence of the Voigt element.

That this is not true can be seen as follows. The Laplace transform of the constitutive relation (1) is

$$\tilde{\tau}_{k\ell} = \lambda(s)\,\tilde{e}_{kk}\delta_{k\ell} + 2\mu(s)\tilde{e}_{k\ell} \qquad (5b)$$

where s is the Laplace transform variable, the tilda represents dependence upon s, and $\lambda(s)$ and $\mu(s)$ are the compliances

$$\lambda(s) = \Big\{\lambda_1 s^2 + \frac{\mu_1\mu_2}{\nu_1\nu_2}(\lambda_1 + \tfrac{2}{3}\mu_1) + \{|\frac{(\mu_1+\mu_2)}{\nu_2} + \frac{\mu_1}{\nu_1}|\,|\lambda_1 + \tfrac{2}{3}\mu_1| - \frac{2}{3}\frac{\mu_1\mu_2}{\nu_2}\}s\Big\} \div (s^2 + |\frac{(\mu_1+\mu_2)}{\nu_2} + \frac{\mu_1}{\nu_1}|s + \frac{\mu_1\mu_2}{\nu_1\nu_2}) \qquad (6a)$$

$$\mu(s) = \frac{\mu_1 s(s + \mu_2/\nu_2)}{s^2 + |(\frac{\mu_1 \quad \mu_2}{\nu_2}) + \frac{\mu_1}{\nu_1}|s + \frac{\mu_1\mu_2}{\nu_1\nu_2}} \qquad (6b)$$

for the simple Burger's body rheology. As shown in Wu and Peltier (1981a), the relaxation spectrum for a spherical, incompressible, homogeneous, visco-elastic Earth model is given by the solution to the algebraic equation

$$\mu(s) = \frac{-\rho_o \ell g a}{(2\ell^2 + 4\ell + 3)} = -\beta \qquad (7)$$

where ρ_o is the density of the planet and a is its radius, g is the surface gravitational acceleration, and ℓ the degree of a spherical harmonic deformation of its surface. For the Burger's body $\mu(s)$ is given by (6b), and (7) is thus a quadratic equation in s which has the following explicit form

$$(\mu_1 + \beta)s^2 + |\frac{\mu_1\mu_2}{\nu_2} + \frac{\beta(\mu_1+\mu_2)}{\nu_2}|s + \frac{\beta\mu_1\mu_2}{\nu_1\nu_2} = 0 \qquad (8)$$

The two roots of (8), $s_{1,2}$, are the inverse relaxation times for the two modes of viscous gravitational relaxation for each value of ℓ which are supported by the Burger's body model. In the limit $\nu_2 \to \infty$ equation (8) reduces to

$$s = \frac{-1}{\nu_1\,[\frac{1}{\beta} - \frac{1}{\mu}]} = s_m \qquad (9)$$

the inverse relaxation time for the Maxwell sphere. It will be clear by inspection of $\tau_m = 1/s_m$ that when $\beta^{-1} >> \mu_1^{-1}$, τ_m is exactly the relaxation time for a degree ℓ surface deformation of a Newtonian viscous sphere (Peltier, 1974). In order to compare the modes of the Burger's body $s_{1,2}$ to that of the Maxwell solid s_m we plot them in Figure 2 as a function of the viscosity contrast ν_2/ν_1. On this Figure we in fact show $s_{1,2}/s_m$ so that each mode of the Burger's body is normalized with respect to the single mode for the corresponding Maxwell solid. In order to compare the models we have also to fix μ_2 for the Burger's body; in Figure 2 we show the dependence of inverse relaxation time on viscosity contrast for several values of the ratio μ_2/μ_1. The choice $\mu_2/\mu_1 \simeq 1$ was favoured by Anderson and Minster (1979) and this choice yields a relaxed modulus which is one-half the unrelaxed value.

Inspection of Figure 2 shows that for $\nu_2 >> \nu_1$ the Burger's body behaves in relaxation as a Maxwell solid with one root $s_1/s_m = 1$ while the other vanishes. It is, however, the opposite limit which concerns us since the Burger's body which fits seismic Q has $\nu_2 << \nu_1$. At this extreme $s_2/s_m \sim 1$ and $s_1/s_m \to \infty$ and therefore even in this limit the Burger's body will respond to surface loading as a Maxwell solid. Only for $\nu_1/\nu_2 > 0(10)$ are both modes simultaneously important and this conclusion is independent of degree ℓ, roots for two values of which are shown in Figure 2. On the basis of this analysis it seems clear that a Burger's body rheology with $\nu_2/\nu_1 << 1$ will be able to simultaneously fit free oscillations and postglacial rebound data. The free oscillations problem for the single relaxation mechanism and the absorption spectrum models will be discussed elsewhere (Yuen and Peltier, 1981).

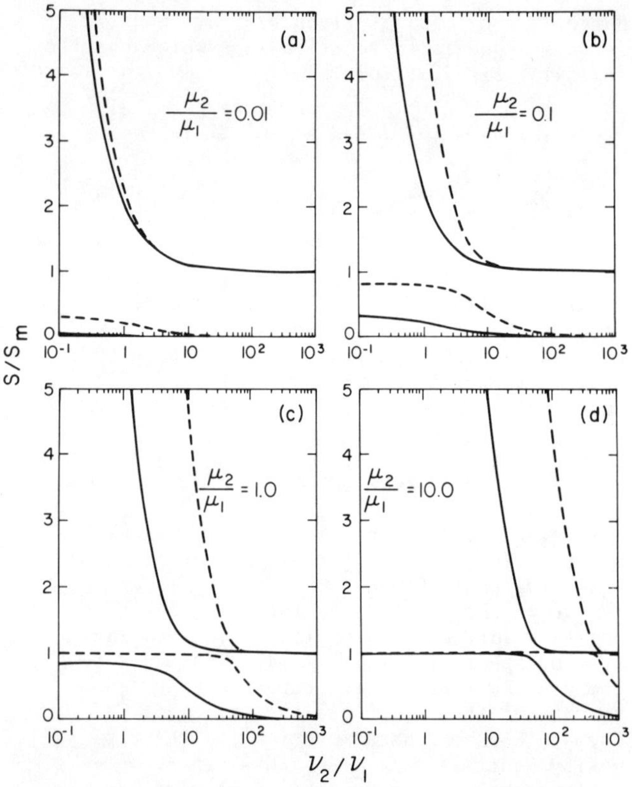

Fig. 2 Inverse relaxation times $s_{1,2}$ of the Burger's body solid as a function of the ratio ν_2/ν_1 for several values of the ratio μ_2/μ_1 as shown. The inverse relaxation times for the two modes of the Burger's body are normalized by that for the Maxwell solid s_m. In each plate the solid lines are the modes for ℓ = 2 and the dashed lines are the modes for ℓ = 50.

Postglacial rebound and transient rheology

The question of whether or not transient rheology could be important in glacial isostasy has recently been raised by Weertman (1978) who revived the idea originally suggested by Goetze (1971). This notion has been further reinforced by Anderson and Minster (1979) based upon inferences from the apparent frequency dependence of seismic Q and they have been led to suggest an explicit model for the transient behaviour which they claim persists for timescales on the order of those which govern isostasy. Other models for transient rheology also exist, the best known of which is probably that described by Garofalo (1965) and which was recently employed by Post (1977) in his analysis of transient creep data obtained in the laboratory. Amin et al. (1970) have provided some theoretical justification for this model based upon unimolecular kinetics. Since these models are both based upon viscosities which have been parameterized with respect to time it does not make sense in terms of them to employ a Burger's body type analogue for the anelastic behaviour. Rather what one does is to choose either the Maxwell model or the standard linear model and to make its viscosity variable, recognizing that this is simply a device intended to mimic the physical effects of transience. Anderson and Minster (1979) use a standard linear solid in which the viscosity is

$$\nu_2(t) = k_o + k_1 t^{2/3} \qquad (10)$$

with $k_1 = 1.08 \times 10^{15}$ Poise/(sec)$^{2/3}$ and $k_o = k_1 t^{2/3}$ with t = 1 year has been introduced in order to remove the unphysical behaviour (zero viscosity) which would otherwise occur at t = 0. The Garofalo type model is based upon a Maxwell analogue with viscosity

$$\nu_1(t) = \frac{\nu_\infty}{(\frac{\nu_\infty}{\nu_o} - 1)\exp(-t/\sigma) + 1} \qquad (11)$$

where σ is a rate constant which governs the timescale over which the viscosity changes from ν_o to ν_∞. The constitutive relations for these solids are respectively (3) with $\nu_2(t)$ as in (10) and (2) with $\nu_1(t)$ as in (11). Our purpose in the present section is to demonstrate that, insofar as postglacial rebound is concerned, neither of the transient rheologies (10) or (11) is compatible with the data.

In order to establish this result we need to develop a method for solving surface loading problems for Earth models with time dependent viscosity. There is in fact a straightforward method for doing this which was described in Peltier et al. (1980) and the following will review the analysis given there. Multiplying (3) by $\nu_2(t)$ from (10) it becomes

$$\nu_2(t)\dot{\tau}_{k\ell} + (\mu_1 + \mu_2)\left(\tau_{k\ell} - \frac{1}{3}\tau_{kk}\delta_{k\ell}\right)$$

$$= 2\mu_1\nu_2(t)\dot{e}_{k\ell} + \lambda_1\nu_2(t)\dot{e}_{kk}\delta_{k\ell}$$

$$+ 2\mu_1\mu_2\left(e_{k\ell} - \frac{1}{3}e_{kk}\delta_{k\ell}\right) \qquad (12)$$

Note in (12) that every term which contains the time derivative of a tensor is multiplied by $\nu_2(t)$. Because this constitutive relation has time dependent coefficients the loading problem cannot be solved by application of the Correspondence Principle--i.e. direct Laplace transformation. If, however, we could transform the problem from the domain of the real physical time t to the domain of a new time τ such that (12) again had constant coefficients then we could again employ Laplace transform methods in τ space. This can clearly be accomplished by defining τ_2 as

$$\nu_2(t)\frac{d}{dt} = \nu_2(t)\frac{d\tau_2}{dt}\,\frac{d}{d\tau_2} \tag{13}$$

and choosing τ_2 such that

$$\nu_2(t)\,\frac{d\tau_2}{dt} = \nu_2' \tag{14}$$

where ν_2' is a constant. From (14) it follows that the new timescale τ_2 is

$$\tau_2(t) = \nu_2' \int_0^t \frac{dt'}{\nu_2(t')} + c \tag{15}$$

For the Anderson and Minster rheology substitution of (10) in (15) gives

$$\tau_2(t) = \frac{3\nu_2'}{k_1}\left[t^{1/3} - \left(\frac{k_o}{k_1}\right)^{1/2}\tan^{-1}\left(\sqrt{\frac{k_1}{k_o}}\,t^{1/3}\right)\right] \tag{16}$$

The same technique may be employed to render the coefficients of (2) with $\nu_1(t)$ as in (11) independent of time. In this case the new timescale τ_1 is

$$\tau_1(t) = t + \sigma\left(1 - \frac{\nu_\infty}{\nu_o}\right)\left(e^{-t/\sigma} - 1\right) \tag{17}$$

where the arbitrary constant has been selected to give $\tau = t$ in the limit $t \to 0$. On the "stretched" timescale the constitutive relations for both transient rheologies now have constant coefficients and so we may solve the loading problem using Correspondence Principle methods.

To compare the rebound predictions of transient models with observed data we should employ the above mapping idea in conjunction with the full gravitationally self-consistent model for relative sea level variations--some new results from the application of which will be discussed in the next section. We can, however, do some meaningful comparisons without introducing the full model if we can isolate information on the variation of relaxation time with time for a single harmonic constituent of the deformation. We could then compare the observed relaxation with the analytic prediction for the relaxation spectrum of a homogeneous, incompressible visco-elastic model. In Wu and Peltier (1981 a) it is shown that the standard linear and Maxwell models have relaxation time spectra which are respectively

$$\tau_2^{\ell} = \nu_2'\,\frac{\left(1 + \dfrac{\rho_o \ell g a}{\mu_1(2\ell^2 + 4\ell + 3)}\right)}{\left(\mu_1 + \dfrac{2\rho_o \ell g a}{(2\ell^2 + 4\ell + 3)}\right)} \tag{18}$$

$$\tau_1^{\ell} = \nu_1'\left(\frac{2\ell^2 + 4\ell + 3}{\ell \rho_o g a} + \frac{1}{\mu_1}\right) \tag{19}$$

where in writing (18) as before we have assumed that the relaxed shear modulus is one half the unrelaxed shear modulus. To transform the exponential relaxation predicted by either model in the τ domain back to the domain of the physical time t we simply note that

$$\exp(-\tau/\tau_{1,2}) = \exp\left(\frac{-t}{\tau_{1,2}}\cdot\frac{\tau(t)}{t}\right) \tag{20}$$

where $\tau(t)$ is the appropriate map (16) or (17) for the standard linear and Maxwell transient cases respectively.

To compare the predictions (20) with real data we will use the relative sea level curve from the Richmond Gulf of Hudson Bay which is near the centre of what was the Laurentide ice sheet. Since the spatial scale of the load was dominated by $\ell = 6$, the Richmond Gulf curve should provide a good approximation to the free decay history of this harmonic. A semi-log plot of the time dependent RSL data from the Richmond Gulf is shown in Figure 3 (after Hilaire-Marcel and Fairbridge, 1978) where we have divided the data into two regions by a line at 5000 years BP. Prior to 5000 years BP nearby regions were undergoing active deglaciation and so only the data subsequent to

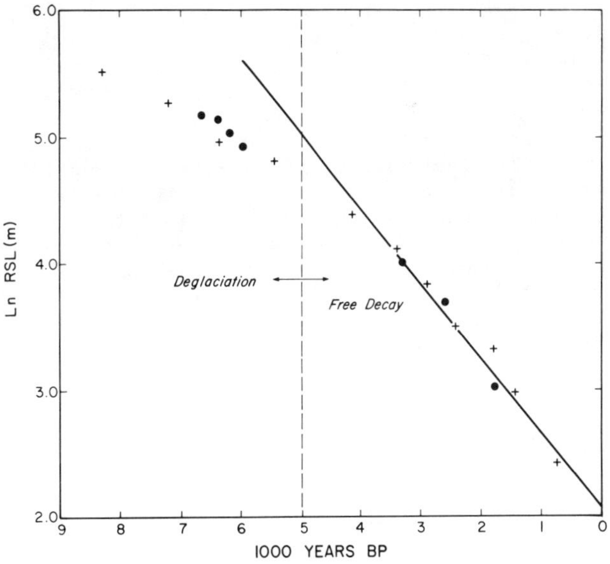

Fig. 3 Relative sea level data from Richmond Gulf in Hudson Bay (solid circles and crosses) after Hilaire-Marcel and Fairbridge (1978). The solid line is a least squares fit to the data in the free decay regime for times later than 5KBP. The age data are all from shells and those shown as crosses have been corrected for isotopic fractionation.

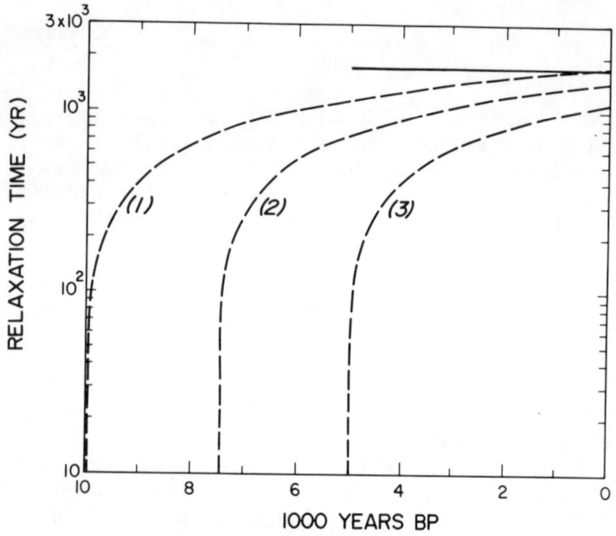

Fig. 4 Relaxation time for a transient standard linear solid rheology as a function of time before present. The solid line represents the free decay regime observations shown in Figure 3. The dashed curves illustrate the temporal evolution of the relaxation time for $\ell = 6$ for different onset times of relaxation as (1) 10^4 yrs BP, (2) 7.5×10^3 yrs BP, (3) 5×10^3 yrs BP.

this time are representative of free decay. The best fit straight line through the data in the free decay regime has a slope which gives a free decay time for $\ell = 6$ of T = 1714 years. If we simply substitute this observed decay time for the ℓ.h.s. of (19) and solve for ν_1' we get a viscosity of $\sim 10^{22}$ Poise for the upper mantle and the simple steady state Maxwell model fits the data nicely and gives the conventional value for viscosity. On the basis of the data shown in Figure 3 we see that there is no evidence at all for any variation of relaxation time with time and therefore no direct support for the necessity of invoking a transient model. The question is, can a transient model fit the data equally well.

In Figure 4 we show comparisons between the relaxation time data for Richmond Gulf and the relaxation times (dashed curve) calculated using the transient creep model of Anderson and Minster (1979). Curves 1, 2, and 3 were computed for different onset times of relaxation using (16), (18), and (20). By inspection of this Figure it is clear that only for onset times of relaxation greater than 10^4 years before present can the transient standard linear model come at all close to fitting the constant relaxation time with time data in the time window 0-5KBP. Even in this case, however, a change of relaxation time by almost a factor of two is predicted and this is not observed. Since the onset time should correspond to the time of most rapid deglaciation and since this was centred on 7.5 KBP (Peltier et al., 1978) the discrepency between this theory and observation is very much larger than a factor of 2 and completely unacceptable. This transient model is strongly rejected by the observations. In order to fit the data a transient model must be able to deliver a constant relaxation time over a rather broad time window.

In Figure 5 we plot τ/t for the transient Maxwell model of Garofalo. The ratio τ/t from (20) can be identified as a decay amplification rate. In the upper part of the Figure we have kept $\nu_\infty/\nu_0 = 10^2$ fixed and varied the rate constant σ. Such models can clearly fit the rebound data (constant τ/t) either if the rate constant is very small, in which case rebound sees the steady state viscosity anyway, or if σ is very large, in which case rebound sees only the initial viscosity and convection would therefore be governed by a viscosity which is two orders of magnitude higher than the one rebound sees. Post's preferred value of σ = 5000 years is intermediate and in the lower part of Figure 4 we show the variation of τ/t with τ fixed at this number and vary ν_∞/ν_0. In order to fit the rebound data with this model we are obliged to accept a very large contrast between the initial and final viscosity values which means that convection would see a viscosity which is several orders of magnitude larger than that which governs rebound. But this would make it difficult to explain plate tectonics with any reasonable version of the convection hypothesis and this hypothesis fits the observations very well if the steady state rebound value of ν_1 is employed. It is extremely unlikely that transient rheology plays any important role in postglacial rebound.

The free air gravity anomaly as a constraint on deep mantle viscosity

Having established in the last section that transient effects are not liable to be important in postglacial rebound, we shall return here to employ the steady state rheological model to infer the viscosity of the deep interior of the planet using postglacial rebound data. Furthermore, the analysis of the Burger's body model presented in Section 2 has demonstrated clearly that this model behaves exactly as a Maxwell solid with viscosity ν_1 insofar as the surface loading problem is concerned, even though the short timescale viscosity ν_2 is much less than the long timescale viscosity ν_1. As stated above, the relation $\nu_2 << \nu_1$ follows from fitting the Burger's body model to the Q's of the normal modes of elastic gravitational free oscillation with ν_1 fixed by postglacial rebound data. This shows that, insofar as the interpretation of rebound data is concerned, the steady state Maxwell rheology suffices.

Over the past several years a gravitationally self-consistent global model of glacial isostatic adjustment has been developed as a vehicle for the inference of mantle viscosity and this model is based upon the assumption that for deformations on the timescale of rebound the Earth behaves as a Maxwell solid. The prediction of relative sea level variations with the model involves the solution of an integral equation which I have previously referred to as the "sea level equation" (Farrell and Clark, 1976; Peltier et al., 1978). If $S(\theta,\phi,t)$ is the relative sea level at any point in the oceans, where relative sea level is the height above present day sea level ($\pm$) of the beach of age t at the location (θ,ϕ), then it may be deduced by solving

$$S(\Theta,\phi,t) = \rho_I \frac{\phi^H}{g} \overset{*}{_I} L_I + \rho_w \frac{\phi^H}{g} \overset{*}{_o} S + C(t) \tag{21}$$

where
$$C(t) = \frac{-M_I(t)}{\rho_w Ao} - \frac{1}{Ao}\left\langle \rho_I \frac{\phi^H}{g} \overset{*}{_I} L_I + \rho_w \frac{\phi^H}{g} \overset{*}{_o} S \right\rangle_o$$

in which ρ_I and ρ_w are respectively the densities of ice and water, g is the gravitational acceleration at the Earth's surface, ϕ^H is the Green's function for the change in surface gravitational potential forced by the addition to the surface of the visco-elastic planet of a unit point mass load (Peltier, 1974, 1976). The functional $L_I(\Theta,\phi,t)$ is the deglaciation history, from which $M_I(t)$ may be computed. $M_I(t)$ is the time dependent loss of mass by the ice sheets due to melting (deglaciation). Ao is the surface area of the oceans so that the first term in the definition of C(t) is just the conventional "eustatic" sea level history. In (21) the operations $\overset{*}{_I}$ and $\overset{*}{_0}$ indicate convolution over the ice and oceans respectively and the symbol $\langle\ \rangle_o$ in the definition of C(t) represents an integral over the oceans. Given the deglaciation history $L_I(\Theta,\phi,t)$, and thus $M_I(t)$, and a visco-elastic Earth model represented by the Green's function $\phi^H(\Theta,\phi,t)$, we may deduce the history of relative sea level $S(\Theta,\phi,t)$ at any point on the surface (Θ,ϕ) by inverting the integral equation (21).

In order to solve this equation we are obliged to discretize our description of the phenomenon, and what is done in practise is to divide the ocean surface and that fraction of the continental surface at which actual deglaciation occurs into a large number of finite elements. We also discretize the system in time and allow simultaneously for the fact that neither deglaciation nor ocean fill-

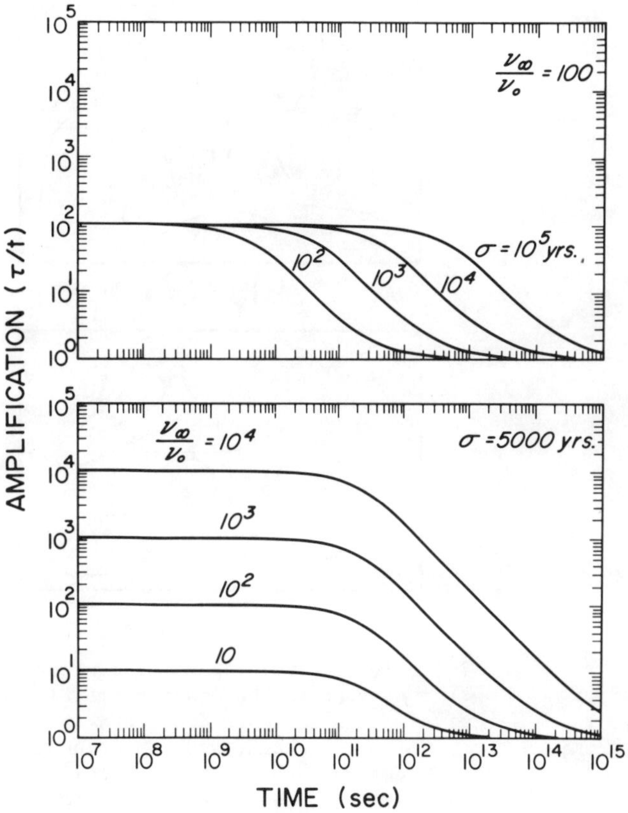

Fig. 5 Amplification factor τ/t for the transient Maxwell model as a function of time. In the top plate the ratio $\nu_\infty/\nu_o = 10^2$ is held fixed while the rate constant σ is varied from 10^2 to 10^5 years. In the bottom plate the rate constant $\sigma = 5 \times 10^3$ years is held fixed while ν_∞/ν_o is varied from 10 to 10^4.

ing are instantaneous. The mass load on the finite element with centroid $\underline{r}'$ is described for either water or ice by

$$L(\underline{r}',t) = \sum_{\ell=1}^{P} L_\ell(\underline{r}')H(t - t_\ell) \tag{22}$$

where $t_\ell(\ell = 1,P)$ is a time series which brackets the loading history at $\underline{r}'$ and $L_\ell(\underline{r}')$ are the loads applied or removed at the discrete times t_ℓ and we assume that the times t_ℓ are common to all active elements. The discrete approximation (22) to the smooth functions $L(\underline{r}',t)$ may be made arbitrarily accurate by choosing $\Delta t = t_{\ell+1} - t_\ell$ sufficiently small. In practice we sample the deglaciation history and sea level response at a uniform $\Delta t = 10^3$ years. In the finite element basis the sea level equation (21) reduces to a set of simultaneous algebraic equations which we solve using a relaxation method al-

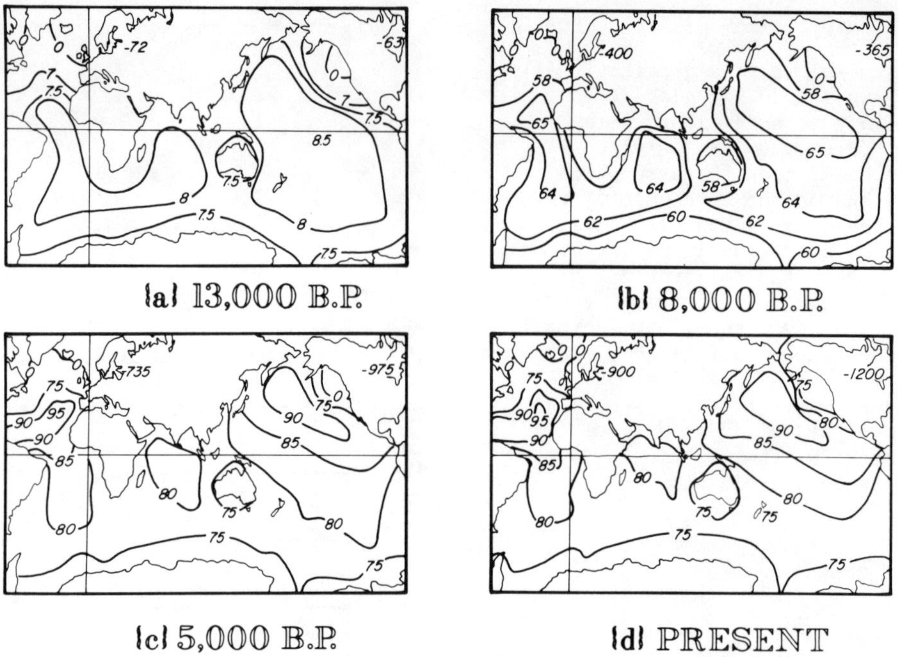

Fig. 6 The global rise of sea level (in metres) at four times subsequent to the onset of melting at 16KBP. Note the large negative values corresponding to a fall of local sea level in the vicinity of the Laurentide and Fennoscandia ice sheets. The rise of sea level is not uniform in the far field showing explicitly that the concept of eustatic sea level is of limited utility.

though a direct method would probably prove preferable. Details of the numerical methods are given in Peltier (1974, 1976), Peltier et al. (1978), and Peltier and Wu (1980 a,b).

An example of a solution to the relative sea level equation (21) is shown in Figure 6 where the global sea level rise in meters is displayed at a sequence of times subsequent to the onset of deglaciation. Inspection of the final frame reveals that the average rise in sea level over the oceans is on the order of 80 metres. Over the Hudson Bay and Fennoscandia there have been respective falls of sea level of 1200 metres and 900 metres. The relative sea level S which one calculates by solving (21) is exactly the information which is obtained in the field by measuring the height above present sea level of each of a sequence of raised or submerged beaches and determining the age of each beach in the sequence by the application of C^{14} dating methods. In order to compare the predictions of theory with observation we simply extract the appropriate time series S(t) for the point (Θ,ϕ) at which we have data. In general the observation points (Θ,ϕ) are not coincident with the centroids of the finite elements into which the surface has been divided and so one must either interpolate the data onto the observation point or use the gravitationally self-consistent loading history implied by the combination of $L_I(\Theta,\phi,t)$ and $S(\Theta,\phi,t)$ and do a new convolution using the potential perturbation Green's function as kernel to predict the sea level history at the passive observation point. In fact the latter method is always employed since it yields a much more accurate result on coastlines which are the source of most RSL data. Other useful information also comes from mid-oceanic islands which, because of their relatively small surface area, actually act like metre sticks attached to the sea floor, insofar as the RSL data which they provide are concerned.

In Figure 7 we show a sequence of six comparisons between theory and observation at a sequence of sites located at varying distances from the centre of the Laurentide ice sheet. The deglaciation history is somewhat modified from that employed to produce Figure 6. The first three locations shown in Figure 7 (Churchill (a), Ottawa Islands (b), Cape Henrietta Maria (c)) are all sites which were once ice covered and located around Hudson Bay near the centre of postglacial emergence. Each of these sites is characterized by a monotonic fall of sea level following deglaciation. The remaining three locations (Boston (d), Clinton Connecticut (e), Bermuda (f)) are sites located outside the ice margin and at increasing distance from the centre. At Boston the

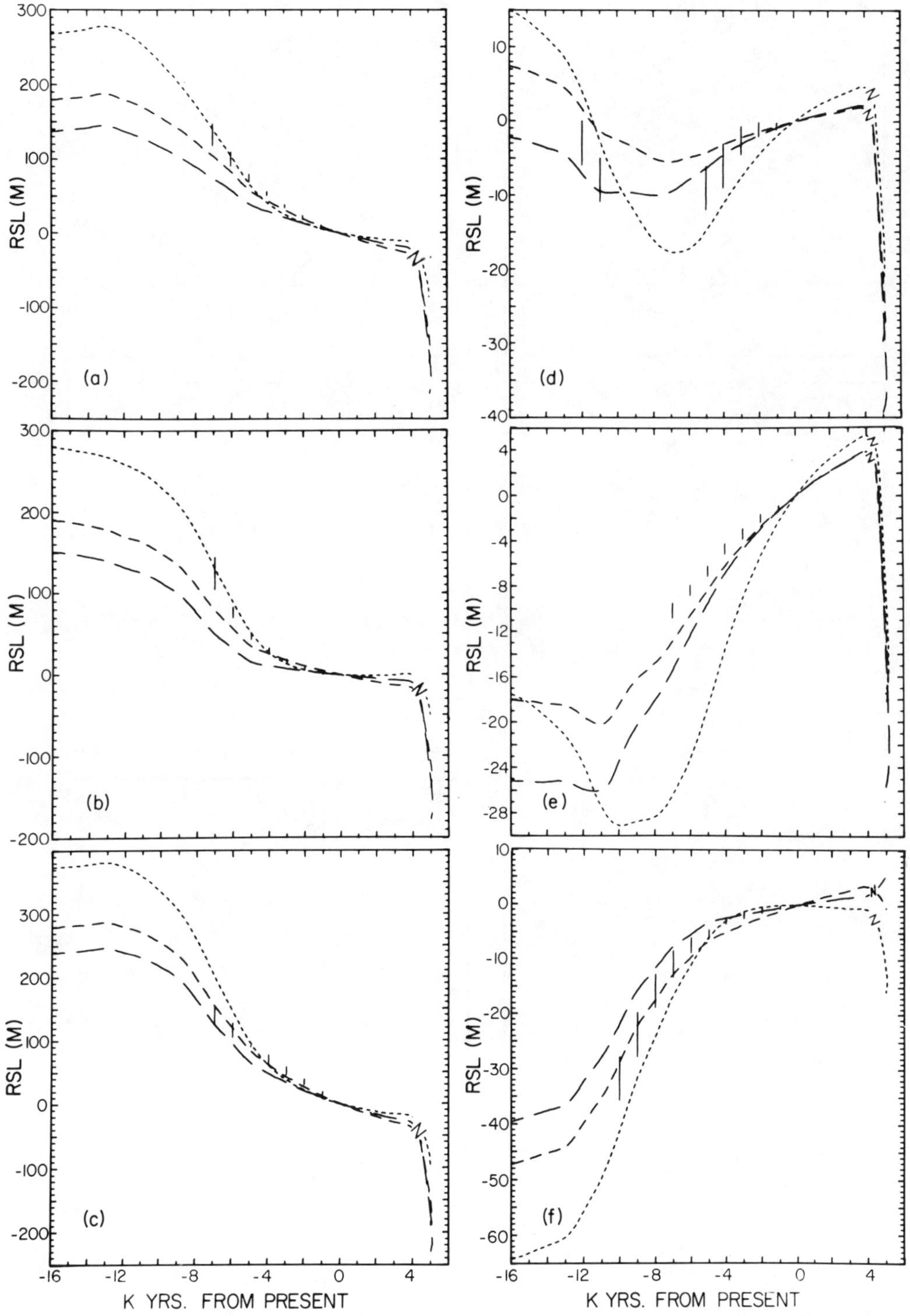

Fig. 7 Comparisons of predicted and observed relative sea level curves for a series of sites in the vicinity of the Laurentide ice sheet. The site locations are (a) Churchill, (b) Ottawa Islands, (c) Cape Henrietta Maria, (d) Boston, (e) Clinton Connecticut, (f) Bermuda. On each plate the dotted, short dashed, and long dashed curves are respectively for the model with uniform 10^{22} Poise mantle viscosity, with a lower mantle viscosity of 10^{23} Poise, and with lower mantle viscosity of 5×10^{23} Poise.

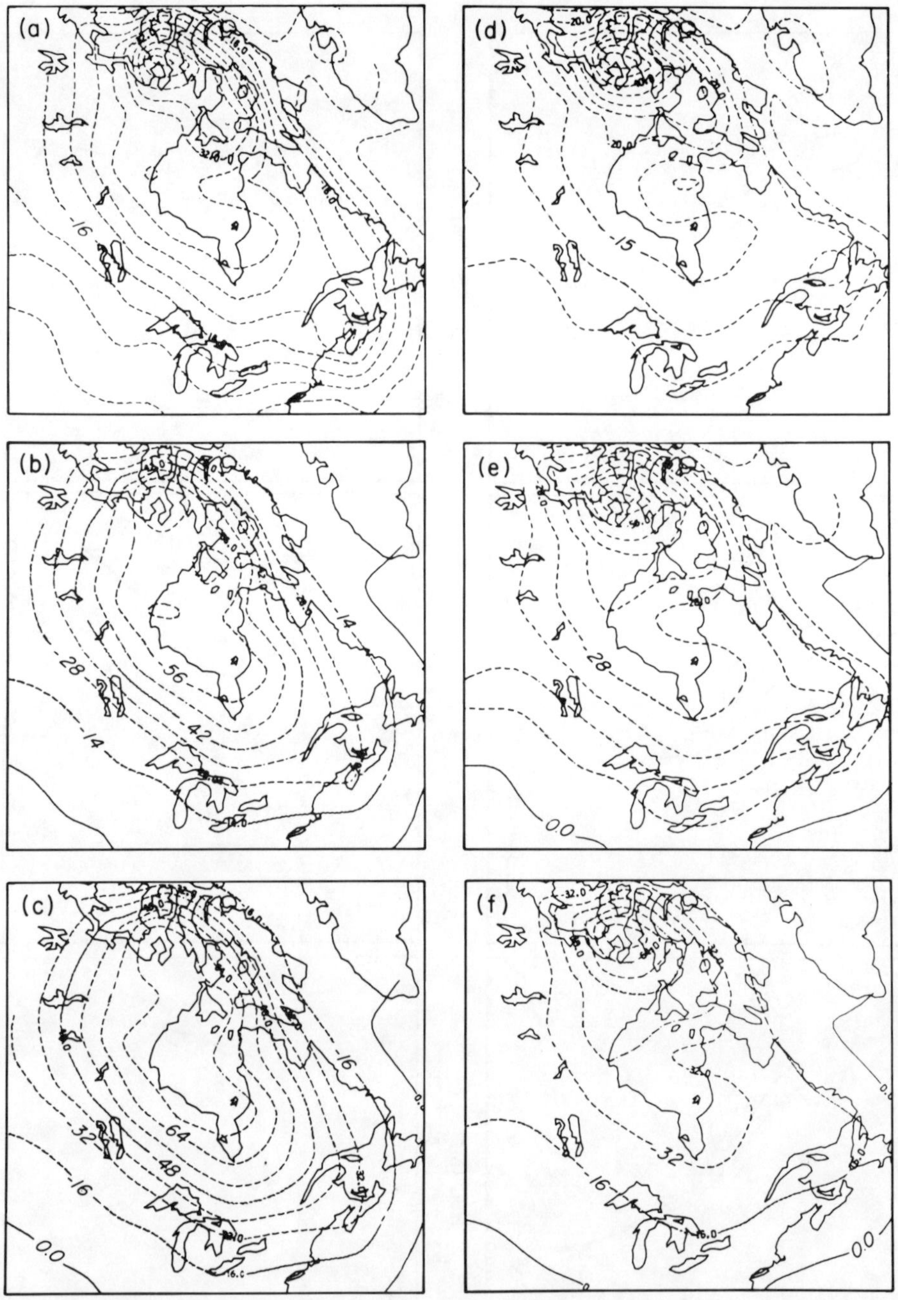

Fig. 8 Theoretically predicted free air gravity maps over the Laurentide depression centred upon Hudson Bay. Plates (a) - (c) are predictions made using the ICE-1 load history whereas (d) - (f) are for ICE-3. As discussed in the text the three viscosity models employed to make each set of predictions differ only in the viscosity of the lower mantle.

rsl data illustrate the characteristic non-monotonic sea level signature of sites nearest the edge of the ice sheet where the propagation of the peripheral bulge (Peltier, 1974) dominates the response. At both of the most distant sites (e and f) the observed rsl variation indicates monotonic submergence. Such response is characteristic of all far field locations except at continental coastlines where tilts induced by the offshore water load cause the development of raised beaches which appear more or less contemporaneous with the cessation of melting (Peltier et al., 1978).

The data from each of these sites are compared

in the Figure with the theoretical predictions from three different visco-elastic models of the mantle. The first model, for which the predictions are shown as dotted lines, has a uniform viscosity of 10^{22} Poise throughout the mantle and a lithosphere which is 120 km thick. The second and third models (short and long dashed lines respectively) differ from model one only in the lower mantle (beginning at a depth of 670 km) where their viscosities are respectively 10^{23} Poise and 5×10^{23} Poise. All three models have $\lambda(r)$, $\mu(r)$, $\rho(r)$ structures equal to those required to fit the periods of the normal modes of elastic gravitational free oscillation (Gilbert and Dziewonski, 1975). It is quite clear by inspection of this Figure that the model with the highest value of the lower mantle viscosity (5×10^{23} Poise) is quite firmly rejected by most of the data. The preferred model is apparently intermediate between the first two and thus an upper bound upon the viscosity of the lower mantle is about 10^{23} Poise. A detailed discussion of such comparisons will be found in Wu and Peltier (1981 b). It is clear by inspection of this limited set of comparisons of theory and observation that the relative sea level data are not themselves enormously sensitive to modest increases of viscosity in the lower mantle. To constrain this quantity more firmly we must invoke other information.

A particularly useful source of additional information is the free air gravity signal over the once ice covered region. To the extent that isostatic disequilibrium currently prevails there should exist negative free air anomalies centred on the areas which were formerly glaciated. Observations of such anomalies provide further constraint upon the viscosity profile in the mantle. Given the self-consistent surface mass load determined by solving the sea level equation (21), a theoretical prediction of the free air map may be made by convolving the load functional with an appropriate Green function. The Green function for the free air anomaly has the form (Peltier, 1979)

$$G^H(\theta, \ ,t) = \left[\frac{g}{m_e} \sum_{\ell=o}^{\infty}\right]$$

$$\times \left[\ell - 2h_\ell^H(t) - (\ell+1)k_\ell^H(t)\right] P_\ell(\cos\theta) \tag{23}$$

where the Love numbers h_ℓ^H and k_ℓ^H each have normal mode expansions of the form

$$h_\ell^H(t) = \sum_j \frac{r_j^\ell}{s_j^\ell}\left(1 - e^{-s_j^\ell t}\right) + h_\ell^E \tag{24}$$

where the s_j^ℓ are the free decay poles in the relaxation spectrum for the decay of the spherical harmonic deformation of degree ℓ, r_j^ℓ are the residues at these poles for a surface point mass forcing, and the h_ℓ^E are the instantaneous elastic contributions to the total response. The gravity map $\Delta g'(\theta,\phi,t)$ may be calculated from the convolution

$$\Delta g(\theta,\phi,t)$$

$$= \sum_i \int\int d\Omega' \, G^H(\theta|\theta', \phi|\phi', t|t_i) \, L_i(\theta',\phi') \tag{25}$$

and the definition $\Delta g' = \Delta g(\theta,\phi,t) - \Delta g(\theta,\phi,\infty)$. In (25) the convolution in time has been replaced by a sum of properly phased and weighted responses to step load removals. The main difficulty in this calculation is the determination of (the isostatic anomaly) $\Delta g(\theta,\phi,\infty)$ and this has been discussed in detail in Wu and Peltier (1981 a).

In Figure 8 is shown a sequence of calculations of the free air anomaly for the Hudson Bay region. The observations for this area were compiled by Walcott (1970), whose map clearly reveals an ellipsoidal negative anomaly of approximately -35 mgals (at minimum) which is very well correlated with the location of the Laurentide ice sheet. The predicted present day anomaly patterns shown in plates (a), (b), and (c) are respectively for the model with uniform 10^{22}P mantle viscosity, with lower mantle viscosity of 10^{23}P, and with lower mantle viscosity of 5×10^{23}P. These three calculations were all done using the ICE-1 deglaciation history tabulated in Peltier and Andrews (1976). Plates (d), (e), and (f) show $\Delta g'$ for the same three viscosity models but for a slightly different deglaciation history called ICE-3. This is the same history employed for the rsl calculations shown in Figure 7. The main difference between this model and ICE-1 is that the Laurentide ice sheet is thinner and melting is delayed by approximately 2000 years (see Wu and Peltier 1981 a,b for discussion).

Inspection of these theoretical predictions show that if the Peltier and Andrews (1976) ice history best describes the actual deglaciation event then a lower mantle viscosity of even 10^{23} Poise cannot be tolerated, since such a model predicts a present day free air gravity anomaly in excess of 60 mgals and this is well outside the errors associated with the 35 mgal observation (e.g. Hudson Bay is a sedimentary basin, etc.). Further increasing the lower mantle viscosity to 5×10^{23} Poise (plate c) only exacerbates the problem. On the other hand, and again in the context of ICE-1, the uniform 10^{22} Poise mantle model fits the free air gravity observations rather well (plate a). These data, therefore, reinforce the conclusion in Peltier and Andrews (1976), based on fits to the rsl data using ICE-1, that uniform mantle viscosity was quite strongly preferred. This has not been demonstrated previously.

The calculations shown in plates (d) - (f) demonstrate clearly the extent to which the inferred viscosity of the lower mantle depends upon the assumed unloading history. With ICE-3 a considerable increase of viscosity in the lower mantle is required to fit the observed free air signal. Indeed, although the third model with a deep mantle viscosity of 5 x 10^{23} P predicts a gravity signal which is somewhat large (plate f), it is not so much in excess of the observation as to be strongly excluded on this basis alone. The model is strongly rejected by the previously shown rsl comparisons however.

This strong constraint upon the viscosity of the deep mantle has recently been confirmed on the basis of yet another line of geophysical argument in Sabadini and Peltier (1980). There it is shown that a lower mantle viscosity near 10^{23} P also fits the observed true polar wander (TPW) evident in the ILS path for the rotation pole and that this geophysical observation is also unambiguously related to the last deglaciation event. The same inference follows from a similar analysis of the observed non-tidal acceleration of the Earth's rotation on the basis of the demonstration that it too is a product of forcing by Pleistocene deglaciation. As discussed in the next section, the implications of these strong constraints on the viscosity of the deep mantle insofar as the convection hypothesis is concerned are rather important.

Mantle Viscosity and Convection

Two crucial pieces of information which are required in the mantle convection hypothesis are obtained from the analysis of postglacial rebound data as described in the last section. These are (1) the thickness of the lithosphere at the surface of the planet and (2) the viscosity of the mantle in the sub-lithospheric region. To appreciate why these observations play such a crucial role we may invoke the boundary layer theory for convection at high Rayleigh and Prandtl number which has been employed by Sharpe and Peltier (1978) to demonstrate compatibility of the whole mantle convection hypothesis with geophysical observables. This boundary layer theory, for heated below convection subject to stress free boundary conditions, was first introduced in the context of discussions of mantle convection by Turcotte and Oxburgh (1967) although it has only been employed by them to discuss upper mantle convection. Its application to the understanding of whole mantle convection was first introduced by Sharpe and Peltier (1978, 1979) in their work on the parameterization of mantle convection in thermal history models and extensions of these ideas are given in Peltier (1980 a,b). The boundary layer theory makes the following predictions for the main properties of the circulation in terms of the Rayleigh number Ra:

$$\delta = a_1(\Delta)\ L\ (Rc/Ra)^{1/3} \qquad (24a)$$

$$u = a_2(\Delta)\ Ra^{2/3}\kappa/L \qquad (24b)$$

$$w = a_3(\Delta)\ Ra^{2/3}\kappa/L \qquad (24c)$$

$$q = a_4(\Delta)\ (Ra/Rc)^{1/3}K\Delta T/L \qquad (24d)$$

where δ is the thermal boundary layer thickness, u and w are characteristic horizontal and vertical velocities, and q is the surface heat flow. For heated below convection the Rayleigh number Ra is defined as $Ra = g\alpha\Delta T L^3/\kappa\nu'$ where g is the gravitational acceleration, α the coefficient of thermal expansion, L the depth of the layer, κ the thermal diffusivity, and ν' the kinematic viscosity. The kinematic viscosity is equal to the molecular value ν deduced from postglacial rebound divided by the mean density. The constants a_i are functions of the aspect ratio of convection Δ and may be determined by matching the boundary layer and plume solutions to the isothermal Stokes flow which obtains in the cell core. Physically realized convective flows in the laboratory are always characterized by an aspect ratio near 1 and it is for this reason that the hypothesis of upper mantle convection is such a difficult one to defend. For the same reason the whole mantle hypothesis is rather attractive since it immediately explains the mean horizontal plate scale of about 4000 km in terms of an interpretation of the plates as forming the lids of whole mantle convection cells which have aspect ratio one.

On the left hand sides of relations (24) all the qualities δ, u, and q are geophysical observables. The thermal boundary layer thickness δ is well approximated by the lithospheric thickness since the strong dependence of viscosity upon temperature makes the cold surface thermal boundary layer visible as the rheological lithosphere in which the viscosity is extremely high. Rebound data yield a continental lithospheric thickness of approximately 110 km. The horizontal velocity u is directly observable as the relative velocities of surface plates, the mean of which is approximately 4 cm/year. This is based upon the interpretation of the oceanic plates as the thermal boundary layer of the mantle wide convective circulation and the assumption that the plates are perfectly coupled to the mantle beneath. The third of the three observables in the boundary layer theory, the mean surface heat flow q, is of course observed directly (e.g. Pollack and Chapman, 1977).

Equations (24) may be manipulated as described in Peltier (1980 a,b) to eliminate the unknown temperature drop across the layer in two distinct ways. If (24 b) and (24 d) are combined then we can solve the resulting expression for the unknown depth of convection L in terms of observables as

$$L_1 = \frac{a_4^{1/2}}{a_2} \left(\frac{c_p \nu'}{R_c^{1/3} q \alpha g}\right)^{1/2} u. \qquad (25)$$

A second estimate of the depth may be obtained by combining (24 a) and (24 b) to obtain

$$L_2 = \frac{(\delta^2/\kappa)}{a_2 a_1 Rc^{2/3}} u \qquad (26)$$

Clearly equations (25) and (26) provide two independent estimates of the depth of convection L in terms of geophysical observables. If in (25) we substitute $u = 4$ cm yr^{-1} as the mean plate speed, $q = .75$ erg cm^{-2} s^{-1} as the surface heat flow, $\nu = \rho\nu' = 10^{22}$ Poise, $g = 10^3$ cm s^{-2}, $\alpha = 3 \times 10^{-5}$ $^oK^{-1}$, $c_p = 1.2 \times 10^7$ erg gm^{-1} $^oK^{-1}$, and $Rc \simeq 10^3$ then we obtain $L_1 \simeq 8 \times 10^3$ km. If we employ a boundary layer thickness $\delta = 100$ km in (26) with the other parameters fixed then we obtain $L_2 \simeq 3.6 \times 10^3$ km. Both the estimates L_1 and L_2 are on the order of the thickness of the whole mantle. That L_1 is high by a factor of about two can be understood as a consequence of the fact that the heat flow employed in (25) should actually represent an average for the spherical shell since in the shell (mantle) the heat flow in steady state increases geometrically with depth for a heated below flow. Increasing q decreases the prediction L_1. It should be noted that the scaling relations (25) and (26) are useless without the rebound measurements of ν and δ. With these measurements, however, they become powerful tools of inference. To the extent that the heated below model is the appropriate model for discussions of mantle convection these scaling relations quite clearly imply that convection should fill the entire mantle. That the heated below driving mechanism is probably dominant has been argued in Peltier (1980 b) and Jarvis and Peltier (1980 a,b) and was assumed in the thermal history models of Sharpe and Peltier (1978, 1979).

We may go one step further in using the scaling relations (25) and (26) by equating L_1 and L_2 and solving the resulting expression for the viscosity of the mantle to obtain

$$\nu' = \delta^4 q \left(\frac{\alpha g}{c_p \kappa^2}\right) \cdot \frac{1}{a_1^4 a_4^{Rc}} \qquad (27)$$

It should not be surprising that when we substitute previously stated values for quantities on the right hand side we get a mean molecular viscosity for the mantle which is near 10^{22} Poise (closer to 10^{23} Poise with necessary parametric corrections) which is just the value which we obtained from the analysis of postglacial rebound and rotation data. Indeed, we may now state that if the viscosity were much different from the effective viscosity obtained in such studies then the convection hypothesis would not be nearly as attractive as it is. Postglacial rebound must be seeing essentially the steady state viscosity and not merely a transient creep value as argued previously in Section 3.

Conclusions

The analyses of postglacial rebound which have been completed to date clearly establish that a simple model with a steady state linear Maxwell rheology is able to fit the isostatic adjustment data extremely well. Discussion given here, and in Peltier et al. (1980), shows that it is extremely difficult to fit the postglacial rebound data with a model with transient rheology even when the transient effects are weak. The viscosity for the mantle which one obtains from rebound studies is precisely that which is required in the convection hypothesis of continental drift--further reinforcing the notion that postglacial rebound sees the steady state value of the momentum diffusion coefficient. When free air gravity data are combined with RSL information then one may require an increase of viscosity across the 670 km seismic discontinuity to satisfy the two data sets simultaneously. The required increase of viscosity, however, is much less than two orders of magnitude and thus is completely inadequate to confine convection to the upper mantle. The hypothesis of whole mantle convection is strongly preferred as a working model of large scale mantle dynamics and has been shown to fit the geophysical observables. The model which fits the data is one in which the "driving mechanism" for convection is due to heating from below--that is due to the loss of heat from the core of the Earth. The importance of this mode of heating was first established indirectly in the thermal history models of Sharpe and Peltier (1978, 1979). Direct evidence that it is the most important forcing in the system is discussed in Jarvis and Peltier (1980 a,b). This idea is strongly supported by the isotopic data presented in O'Nions et al. (1978). An important consequence of this mode of forcing is that there must exist a sharp thermal boundary layer at the core mantle interface. There is direct seismic evidence for the existence of this region in the form of the D" layer (Cleary, 1974) in which seismic velocities deviate sharply away from their values in a smooth model like PEM-A (Dziewonski,

Hales, and Lapwood, 1975). It has been argued (Peltier 1980 a,b; Yuen and Peltier 1979 a,b) that this boundary layer should be at times violently unstable against secondary convective instability and that such instabilities could manifest themselves as thermal plumes with small spatial scales which are expected to rise quickly to the base of the lithosphere and there induce intense partial melting. These remarks on the important implications of the relatively weak variation of viscosity with depth which is inferred from postglacial rebound studies serve to emphasize the important role which the isostatic adjustment data play in the convection hypothesis.

There have been several arguments given recently to the effect that Newtonian rheologies for the mantle are completely inappropriate and if this were true it would cast in considerable doubt the ideas discussed above. In the remainder of this section we will consider these arguments in turn and show that they do not withstand close inspection.

Post and Griggs (1973) suggested on the basis of the stress strain relation for a non-Newtonian power law fluid in the form $c\dot{e}^{1/n} = \tau$ that the rebound data themselves provided evidence of non-Newtonian flow. In the power law rheology c is called the creep coefficient and n is the power law exponent. Using dimensional arguments they showed that the maximum rate of uplift dz/dt would be proportional to z^n where z is the amount of uplift remaining. Brennan (1974) derived an approximate solution which gave $dz/dt = -(\rho g/cK)^n z^n/k_H$ where K is a numerical constant which is a function of n and k_H is the horizontal wave number of the deformation. If n = 1 then K = 1 and c = ν and this equation describes the relaxation of a Newtonian halfspace. On the basis of Brennan's solution or Post and Griggs prior dimensional result it is clear that if we plot $\log |dz/dt| = \log [(\rho g/cK)^n k_H^{-1}] + n \log z$ as a function of log z then the slope of the straight line will yield the stress exponent n. The difficulty with applying this idea is that it requires us to estimate the present day amount of uplift remaining which is normally accomplished using an estimate of the free air gravity anomaly over the central depression. The smaller the estimate of the present day free air anomaly the smaller will be the estimate of the amount of uplift remaining and consequently the smaller will be the inferred value of the stress exponent n. Post and Griggs (1973) and Crough (1977) assumed an amount of uplift remaining under Fennoscandia of about 180 m based upon early calculations of Vening-Meinesz (1937) which were based upon rather inaccurate gravity data. This led them to estimate $n \simeq 3$ for the stress exponent. Modern gravity data (e.g. Marsh and Marsh, 1976) show a gravity anomaly over Fennoscandia of only about 10 mgal which converts to a remaining uplift of about 56 m and a predicted power law exponent of unity! There is no unambiguous evidence in the postglacial rebound data themselves which requires a non-Newtonian rheology in spite of recent claims to the contrary (e.g. Melosh, 1979). Such claims are motivated by a lack of understanding of the data.

A second argument which has been given as direct support for the non-Newtonian nature of the rheology of the mantle is that due to Melosh (1976). He has argued that it is impossible to understand the characteristic pattern of stress propagation in the lithosphere (following a large plate margin earthquake (the Rat Island event) without invoking non-linear behaviour. Savage and Prescott (1978), however, have shown quite clearly that his conclusions are conditioned by inadequacies of the model employed to invert the data. The Melosh model neglects the instantaneous elasticity of the mantle beneath the lithosphere. When this is included a linear model seems to fit the data as well as the non-linear one despite Melosh's (1978, 1979) claims to the contrary. A second fault of his model is his assumption of an extremely small background stress, one which is at least an order of magnitude smaller than that associated with the mantle convective circulation. In this context, it is interesting to note that Yamashita (1979), using a standard linear solid rheology with $\nu_2 = 10^{17-18}$ Poise, was able to explain postseismic deformation in terms of aftershock occurences. As mentioned previously, this value for the short term viscosity is also obtained by fitting the single relaxation time SLS to the Q's of the

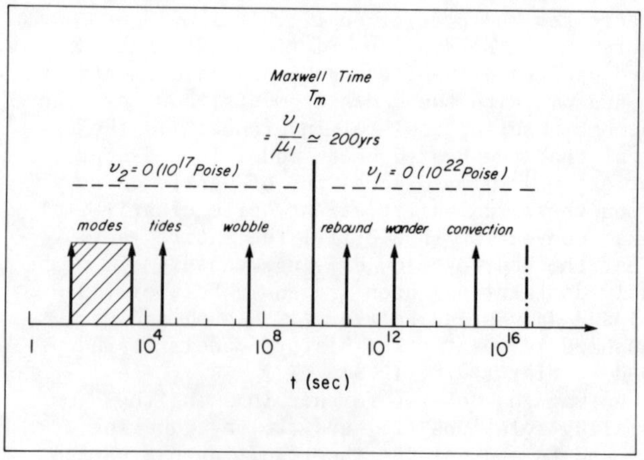

Fig. 9 Diagram illustrating the relation of the characteristic timescale for several geodynamic phenomena to the Maxwell time of the upper mantle. Long timescale phenomena "see" ν_1, while short timescale phenomena "see" ν_2 (or the ν_2 spectrum in the case of an absorption band).

low order fundamental normal modes of free oscillation. There does not seem to be any direct geophysical observation which requires a non-Newtonian rheology to explain it.

Since this is the case, it would appear to be useful to attempt to construct a general linear visco-elastic model and to see if such a model can be employed to reconcile the observations of anelastic behaviour of the Earth. One such model, the simple Burger's body, was discussed in Section 2 where its general constitutive relation was stated. Preliminary evidence suggests that this will in fact prove to be an extremely useful vehicle in terms of which anelastic behaviour might be summarized, particularly when the short timescale anelasticity is described by employing the absorption band model. The long time viscosity ν_1 is fixed by postglacial rebound observations, the short time viscosity ν_2 (or its spectrum for the absorption band model), by observations of seismic Q, while the elastic parameters are determined by the free oscillation frequencies. With these parameters so constrained, how well is the model able to account for other anelastic process such as those involving the Earth's rotation (e.g. the observed speed of true polar wander and the Q of the Chandler wobble) and other geophysical phenomena? In Figure 9 we show the characteristic times for several important geodynamic phenomena. Phenomena with characteristic times shorter than the Maxwell time will be governed, insofar as their anelasticity is concerned, by the short timescale viscosity ν_2 (or the viscosity spectrum for the absorpton band) whereas, when the phenomenological timescale exceeds the Maxwell time, creep is governed by the long-timescale viscosity ν_1. The Maxwell time for the upper mantle is on the order of a few hundred years.

References

Amin, K. E., Mukherjee, A. K., and J. E. Dorn, A universal law for high temperature diffusion controlled transient creep, J. Mech. Phys. Solids, 18, 413 - 426, 1970.

Anderson, Don L., The anelasticity of the mantle, Geophys. J. R. astr. Soc., 14, 135 - 164, 1967.

Anderson, Don L., Kanamori, H., Hart,R. S., and H. P. Liu, The earth as a seismic absorption band, Science, 196, 1104 - 1106, 1976.

Anderson, Don L., and R. S. Hart, Q of the Earth, J. Geophys. Res., 83, 5869 -5882, 1978.

Anderson, Don L., and J. Bernard Minster, The frequency dependence of Q in the Earth and implications for mantle rheology and the Chandler wobble, Geophys. J. R. astr. Soc., 58, 431 - 440, 1979.

Berckhemer, H., Auer, F., and J. Drisler, High temperature anelasticity and elasticity of mantle peridotite, Phys. Earth Planet Int., 20, 48 - 59, 1979.

Brennan, C., Isostatic recovery and strain rate dependent viscosity of the Earth's mantle, J. Geophys. Res., 79, 3993 - 4001, 1974.

Christensen, R. M. Theory of Viscoelasticity: An Introduction, Chapter 2, Academic Press, New York, 1971.

Cleary, J. R., The D'' region at the core-mantle boundary from observations of diffracted S, Phys. Earth Planet. Int., 9, 13 - 27, 1974.

Crough, T., Isostatic rebound and power-law flow in the asthenosphere, Geophys. J. R. astr. Soc., 50, 723 - 738, 1977.

Dziewonski, A. M., Hales, A. L., and E. R. Lapwood, Parametrically simple Earth models consistent with geophysical data, Phys. Earth Planet. Int., 10, 12 -48, 1975.

Farrell, W. E. and J. A. Clark, On postglacial sea level, Geophys. J. R. astr. Soc., 46, 647 - 667, 1976.

Gilbert, F., and A. Dziewonski, An application of normal mode theory to the retrieval of structural parameters and source mechanisms from seismic spectra, Phil. Trans. R. Soc. Lond., A 278, 187 - 269, 1975.

Gittus, J., Creep, Viscoelasticity and Creep-Fracture in Solids, Applied Sci. Ltd., London, 1975.

Goetze, C., High temperature rheology of Westerly granite, J. Geophys. Res., 76, 1223 -1230, 1971.

Goetze, C., and W. F. Brace, Laboratory observations of high temperature rheology of rocks, Tectonophysics, 13, 583 - 600, 1972.

Gross, B., Mathematical Structure of the Theories of Viscoelasticity, Chapter 8, Hermann, Paris, 1953.

Hilaire-Marcel, C., and R. W. Fairbridge, Isostasy and eustasy in Hudson Bay, Geology, 6, 117 - 122, 1978.

Jackson, D. D., and Don L. Anderson, Physical mechanisms of seismic-wave attenuation, Rev. Geophys. Space Phys., 8, 1 - 63, 1970.

Jarvis, G. T., and W. R. Peltier, Flattening of oceanic bathymetry profile due to radiogenic heating in a convecting mantle, Nature, Vol. 285, No. 5767, pp. 649-651, 1980.

Jarvis, G. T., and W. R. Peltier, Mantle convection as a boundary layer phenomenon, Geophys. J.R. astr. Soc., submitted, 1981.

Joncich, D. M., The Plastic Behaviour of predeformed ice crystals, Ph. D. Thesis, Dept. of Physics, Univ. of Illinois, Urbana, 1976.

Kohlstedt, D. L., and C. Goetze, Low stress and high temperature creep in Olivine single crystals, J. Geophys. Res., 79, 2045 - 2051, 1974.

Kohlstedt, D. L., Goetze, C., and W. B. Durham, Experimental deformation of single crystal Olivine with applications to flow in the mantle, in The Physics and Chemistry of Minerals and Rocks, ed. by R. G. Strens, J. Wiley and Sons, Inc., New York, pp. 35 - 50, 1976.

Liu, H. P., Anderson, Don L., and H. Kanamori, Velocity dispersion due to anelasticity:

Implications for seismology and mantle composition, Geophys. J. R. astr. Soc., 47, 41 - 58, 1976.
MacDonald, J. R., Theory and application of a superposition model of internal friction and creep, J. Applied Phys., 32, 2385 -2398, 1961.
Malvern, L. E., Introduction to the mechanics of a continuous medium, Prentice Hall Inc., Englewood Cliffs, New Jersey, 1969.
Marsh, B. D., and J. G. Marsh, On global gravity anomalies and two scale mantle convection, J. Geophys. Res., 81, 5267 - 5280, 1976.
Melosh, H. J., Nonlinear stress propagation in the earth's upper mantle, J. Geophys. Res., 81, 5621 - 5632, 1976.
Melosh, H. J., Reply, J. Geophys. Res., 83, 5009 - 5010, 1978.
Melosh, H. J., Rheology of the earth: theory and observation, in Physics of the Earth's Interior, Enrico Fermi International School of Physics, Course 78, ed. by A. Dziewonski and E. Boschi, North Holland Elsevier Co., 1980.
Mott, N. F., A theory of work hardening of metals, II: flow without slip lines, recovery and creep, Phil. Mag., 44, 742 - 765, 1953.
Murell, S. A. F., and S. Chakravarty, Some new rheological experiments on igneous rocks at temperatures up to 1120°C, Geophys. J. R. astr. Soc., 34, 211 - 250, 1973.
Nowick, A. S. and B. S. Berry, Anelastic Relaxation in Crystalline Solids, Chapter 9, Academic Press, New York, 1972.
O'Nions, R. K., N. M. Evensen, P. J. Hamilton, and S. R. Carter, Melting of the mantle past and present: isotope and trace element evidence, Phil. Trans. R. Soc. Lond. A, 258, 547 - 559, 1978.
Peltier, W. R., The impulse response of a Maxwell Earth, Rev. Geophys. Space Phys., 12, 649 - 669, 1974.
Peltier, W. R., Glacial isostatic adjustment - II. The inverse problem, Geophys. J. R. astr. Soc., 46, 669 - 706, 1976.
Peltier, W. R., Mantle convection and viscosity, in Physics of the Earth's Interior, Enrico Fermi International School of Physics, Course 78, ed. by A. Dziewonski and E. Boschi, North Holland Elsevier Co., 1980 a.
Peltier, W. R., Surface plates and deep mantle plumes: separate scales of the mantle convective circulation, in American Geophys. Union Monograph, ed. by W. Fyfe, E. Lubimova, and R. J. O'Connell, 1980 b.
Peltier, W. R., and J. T. Andrews, Glacial isostatic adjustment - I. The forward problem, Geophys. J. R. astr. Soc., 46, 605 - 646, 1976.
Peltier, W. R., W. E. Farrell, and J. A. Clark, Glacial isostasy and relative sea level: a global finite element model, Tectonophysics, 50, 81 - 110, 1978.
Peltier, W.R., Yuen, D.A., and P. Wu, Postglacial rebound and transient rheology, Geophys. Res. Lett. 7, 10, 733-736,1980.
Pollack, H.N. and D.S. Chapman, Mantle heat flow, Earth Planet. Sci. Lett., 34, 174-186, 1977.
Post, R.L. Jr., High temperature creep of Mt. Burnet dunite, Tectonophysics, 42, 75-110, 1977.
Post, R.L., and D.T. Griggs, The earth's mantle: Evidence of non-Newtonian flow, Science, 181, 1242-1244, 1973.
Sammis, C.G., Smith, J.C., Schubert, G., and D.A. Yuen, The viscosity depth profile of the earth's mantle: effects of polymorphic phase transitions, J. Geophys. Res., 82, 3747-3761, 1977.
Savage, J.C., and W.H. Prescott, Comment on Nonlinear stress propagation in the earth's upper mantle by H.J. Melosh, J. Geophys. Res., 83, 5005-5007, 1978.
Sharpe, H.N., and W.R. Peltier, Parameterized mantle convection and the Earth's thermal history, Geophys. Res. Lett., 5, 737-744, 1978.
Sharpe, H.N., and W.R. Peltier, A thermal history model for the Earth with parameterized convection, Geophys. J.R. astr. Soc., 59, 171-203, 1979.
Turcotte, D.L. and E.R. Oxburgh, Finite amplitude convective cells and continental drift, J. Fluid Mech., 28, 24-42, 1967.
Twiss, R.J., Structural superplastic creep and linear viscosity in the Earth's mantle, Earth and Planet. Sci. Lett., 33, 86-100, 1976.
Vening Meinesz, F.A., The determination of the earth's plasticity from postglacial uplift of Fennoscandia: Isostatic adjustment, Proc. Kon. Ned. Akad. Wetensch, 40, 654-665, 1937.
Walcott, R.I., Isostatic response to loading of the crust in Canada, Can. J. Earth Sci., 7, 716-727, 1970.
Weertman, J., Creep laws for the mantle of the Earth, Phil. Trans. R. Soc. Lond., A, 288, 9-26, 1978.
Wu, P., and W.R. Peltier, Viscous gravitational relaxation, Geophys. J.R. astr. Soc., submitted, 1981 a.
Wu, P. and W.R. Peltier, Glacial isostasy and the free air gravity anomaly as a constraint on lower mantle viscosity, Geophys. J.R. astr. Soc., submitted 1981 b.
Yamashita, T., Aftershock occurence due to viscoelastic stress recovery and an estimate of the tectonic stress field near the San Andreas fault system, Bull. Seism. Soc. Am., 69, 661-688, 1979.
Yuen, D.A., and W.R. Peltier, Temperature dependent viscosity and local instabilities in mantle convection, in Physics of the Earth's Interior, Enrico Fermi International

School of Physics, Course 78, ed. by A. Dziewonski and E. Boschi, North Holland-Elsevier Co., 1980.

Yuen, D.A. and W.R. Peltier, Mantle plumes and the thermal stability of the D'' layer, Geophys. Res. Lett., vol. 7, No. 9, 625-628, 1980.

Yuen, D.A. and W.R. Peltier, Normal modes of the viscoelastic earth, Geophys. J.R. astr. Soc. submitted, 1981.

ROCK MASS CHARACTERISATION BY VELOCITY AND Q MEASUREMENT WITH ULTRASONICS

G.P. Stacey and M.T. Gladwin

Department of Physics, University of Queensland,
Brisbane, Queensland, 4067 Aust.

Abstract. A technique for assessing the blasting characteristics of rock ahead of an active mine face, using velocity and attenuation of ultrasonic pulses propagating between drill holes, has been used on a trial basis in quarries and a mine. Our determination of the mean Q for a rock path is based on measurement of the rise time of the first pulse arrival, a method that has recently been given a rigorous theoretical basis by B.J. Brennan. Information on rock heterogeneity and fracture not available from conventional down-hole techniques is obtained. The pulse rise time method has some advantage over the spectral ratio method of determining Q and appears suitable for upper mantle studies.

Introduction

The viability of a technique for rock mass characterisation using ultrasonic signals has been demonstrated as a part of a programme on rock fragmentation. This programme required development of instrumentation for rapid and easy diagnosis of those physical properties of a rock mass which are relevant to blast design for optimum fragmentation (Just and Walter, 1978). The properties measured were the compressional wave velocity (V_p) and the attenuation coefficient (Q^{-1}). Typical rock masses are heterogeneous so that a discrete transmitted compressional pulse is received as an extended wave train of compressional and shear arrivals from multiple reflections by the numerous cracks and interfaces along the path. This wave train contains information about the cracks and interfaces, but in general the complexity of the received wave form makes inversion to a mapping of the rock mass very difficult (Price, 1975).

The first shear arrival is generally buried in the reflected compressional arrivals, so that, though it has been claimed (T.R. Stacey, 1976) that shear waves are a more sensitive indicator of rock fracture, the first compressional phase is more reliably determined in the field. Because of its dependence on the averaged bulk modulus of the rock mass, the compressional velocity is expected to reflect the microfracture density. In particular, the pressure dependence of velocity at low confining pressure (<200 MPa)(Walsh, 1965; Volarovich et al., 1966) is explained by the change of rock fabric with pressure.

Rock mass characterisation using velocity alone has been attempted previously, but, due to limited success, no system of routine analysis has been developed. Seismic and ultrasonic velocity measurements (Murphy, 1972; Bernabini and Borelli, 1974; Meister, 1974; Grujic, 1974) have been used to delineate heavily fractured rock from competent rock near excavation surfaces. Meister suggested that attenuation data were more useful for rock characterisation, so we have added to the velocity technique a measurement of the attenuation based on the simple scaling law reported by Gladwin and Stacey (1974), who found that a propagating pulse in a linear viscoelastic medium has an intrinsic shape which is merely attenuated and broadened in proportion to the travel time in the medium.

Winkler et al. (1979) have documented the transition from non-linear attenuation in the sonic frequency range. Brennan and Stacey (1977), using cycled torsional strains in the seismic frequency range, found no evidence for non-linearity at strain amplitudes below 10^{-6} at room temperature. These observations allow analysis of attenuation of a complex propagating wave by linear visco-elastic theory, in terms of attenuation of its Fourier components. For pulse propagation, the preferential absorption of the higher frequency components results in an increase in pulse breadth as well as the decrease in amplitude expected from the energy absorption.

The experimental basis of the simple scaling law used in the present technique was first presented by Gladwin and Stacey (1974). Figure 1 shows the definition of pulse rise time, τ used here, which is the inverse of the maximum normalised 'slope' on the displacement pulse,

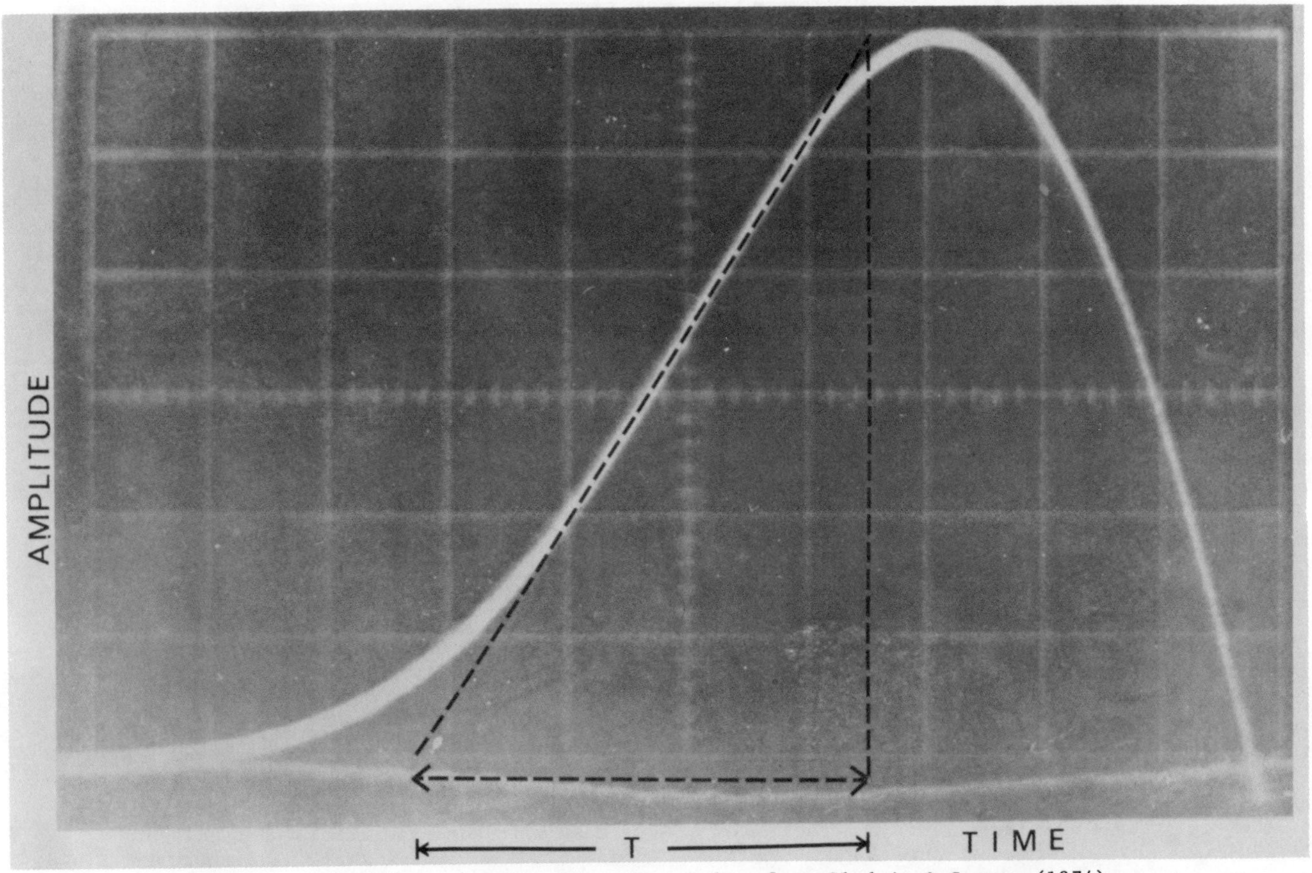

Fig. 1. Definition of pulse rise time, taken from Gladwin & Stacey (1974).

$$\tau = V_{max} / \left(\frac{dV}{dt}\right)_{max}$$

Thus defined, the rise time of a propagating pulse increases from an initial value τ_0 according to a simple law

$$\tau = \tau_0 + C \int_0^T \frac{dt}{Q}$$

for travel over the time $0<t<T$, where Q is the attenuation and C is approximately 0.5. For a path traversing a series of media, each of constant Q, the change of rise time is a sum of the contributions corresponding to the travel time T_i in each layer (i), that is

$$\tau - \tau_0 = C \sum_{i=1}^{N} \frac{T_i}{Q_i}$$

This scaling law was first presented as an empirical result by Gladwin and Stacey (1974), and it has been recently derived by Kjartansson (1979) on the basis of a logarithmic creep law, and also, as a limit case, by Brennan (1980, 1981). If Q is independent of frequency, C=0.485, and Brennan (1980) discusses the generalisation of the law to arbitrary frequency dependence of Q. The theoretical value for constant Q agrees reasonably with the empirical value (0.53 ± 0.04) found by Gladwin and Stacey (1974). This value refers to the particle displacement waveform. Kjartansson (1979) also derived the rise time parameter for a velocity pulse, $C_v = 0.298$, allowing direct use of the rise time law with velocity seismograms. The pulse rise time technique has certain advantages over the direct spectral ratio methods, used with some success by Obert et al. (1976) for characterisation of rock mass. Only a short record is required, and the measurement is less subject to interference by later arrivals which often distort Fourier spectra by violation of the stationarity requirement over the window lengths used to determine spectra.

Instrument

A piezomagnetic transducer is used for a pulse source, and the time of flight is measured

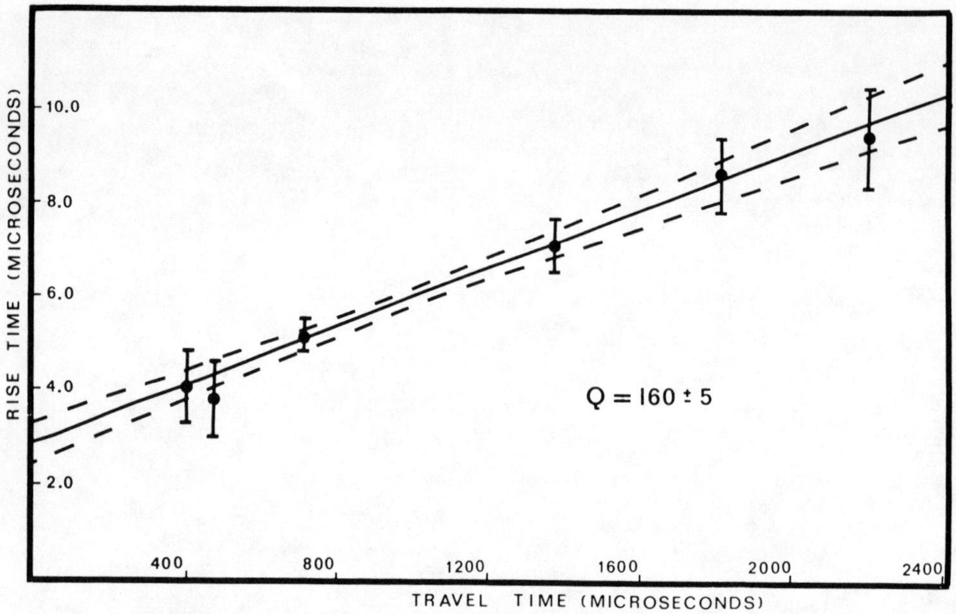

Fig. 2. Pulse rise time versus travel time for ultrasonic pulses propagating in relatively unfractured rock, between boreholes in the floor of a quarry at Ashgrove, Queensland.

between two broadband piezoelectric detectors in the manner previously described by Gladwin (1973). All transducers are screw clamped at any depth in test holes, allowing a complete log of the test area to be taken. Minimum test hole size is 57 mm diameter. For simple velocity/attenuation scans in which the distance between the holes is only surveyed to approximately 2%, both velocity and attenuation measurements can be made with adequate precision by direct

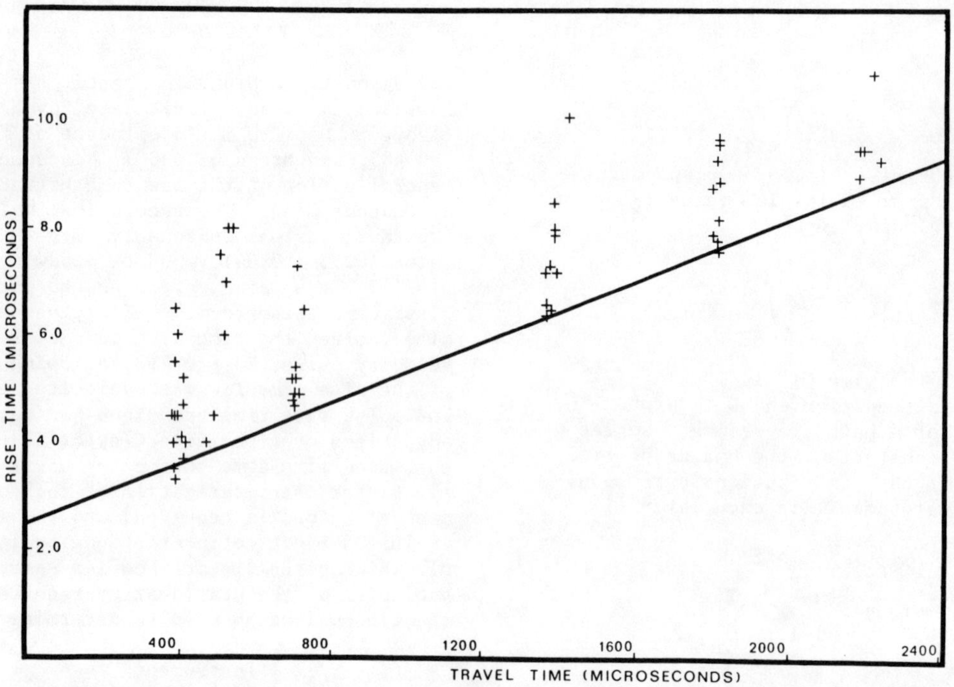

Fig. 3. Pulse rise time versus travel time for all Ashgrove data, including highly fractured paths. The gradient of the line joining the shortest rise times gives the highest Q path available. Points above this line indicate measurement paths in fractured material.

observation on a delayed-time-base oscilloscope. In some applications, higher precision is required to monitor changes in microfracture density associated with a recovery sequence. In these cases, a precise timing module (not shown) is used in conjunction with a digital storage oscilloscope for the Q measurement. The timing module produces averages over 10^4 pulse transits at precision approaching 10 nanoseconds. The storage oscilloscope (a Nicolette Explorer) is used also to sum over the many pulses to improve signal-to-noise ratio on the rise time measurements. In a suite of measurements, changes of Q can be determined to 2%, changes of velocity to 0.01%. The technique is limited to reasonably competent rock with Q greater than about 25 in which distances up to about 30 m are achievable. In most mining environments this gives reasonable capability. The range can be extended by use of lower frequency sources, or by making the measurements on small blasts. One-kilogram charges of Mollenite have been used successfully out to 70 m. There is no apparent limitation on the use of lower frequencies.

Results

A typical set of results from a test site at a quarry at Ashgrove near Brisbane is summarised in Figure 2. The data illustrate the rise time law itself, and the slope is used to determine an average Q of the rock ($\bar{Q}$ = 160 ± 5). A better indication of the usefulness of the Q measurement is shown in Figure 3, in which all rise time measurements for paths between a set of five holes are shown. The lower bound (shown lined) represents the highest Q data set across the region and provides a best estimate of the undisturbed Q of the region (172 ± 6). Increased attenuation at any distance or depth scatters the rise time data towards higher values (lower Q) giving a good estimate of the variability of the attenuation with depth for each path.

A general correlation between travel time and Q is observed at the sites investigated. The change of character of the rock is generally more obvious in the rise time data than in travel time data. At one test site, velocity anisotropy of 10% was found to be associated with Q anisotropy up to 50% correlated with the fabric, high velocity and high Q being observed in the plane of layering. The information from pulse rise time profiles has been compared with structural information extrapolated from nearby cored holes. An example is shown in figure 4.

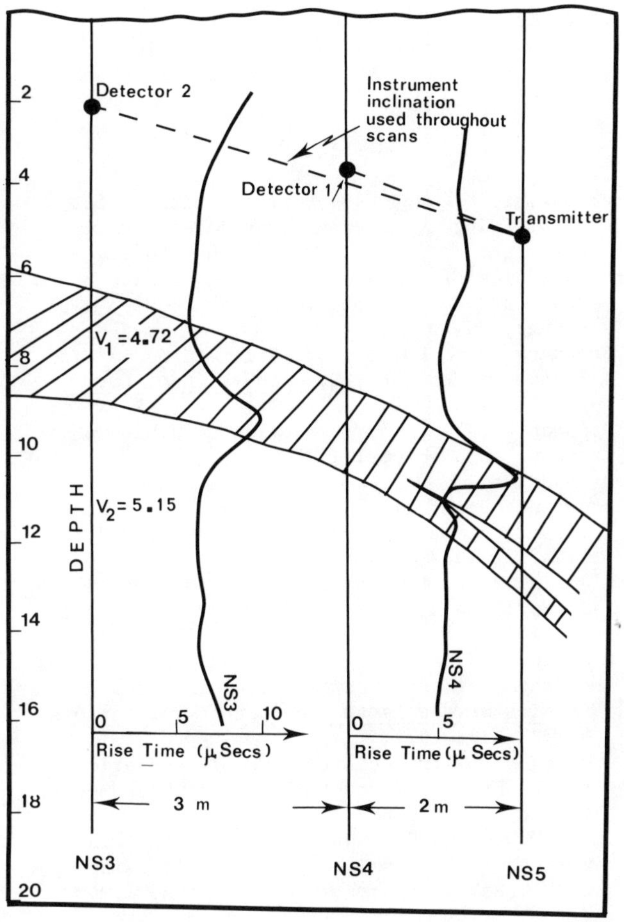

Fig. 4. Comparison of pulse rise time data with cored log for typical site. The shaded area represents a band of thermally altered granite. Transmitter was located in hole NS5. Rise times to inclined detectors in holes NS4 and NS3 are shown. The shaded low Q region is well mapped by the ultrasonic technique.

Discussion

The success of the pulse rise time measurement in providing information on rock character which is not available through velocity techniques alone, strongly suggests that the technique can be applied in the more fundamental problem of crustal rock characterisation. Kjartansson has shown that the rise time technique is very useful for successful prediction of seismic pulse arrivals in a multilayered medium. The extension of the theoretical basis of the technique (Brennan, 1980,1981) to frequency-dependent Q environments should allow some Q estimates for the Earth. Pulse onset techniques have an obvious advantage over spectral ratio techniques in that a significantly shorter record is required. In quarry or mine situations, the lack of resolution in first arrival data can be improved by multiple sampling, and some advantage may be obtained also in seismology by summing arrivals which have traversed similar paths. Limitations on spectral ratio techniques imposed by multiple arrivals are greatly reduced in the pulse rise time method for estimating Q.

References

Bernabini, M., and G.B. Borelli, Methods for determining the average dynamic elastic properties of a fractured rock mass and the variation of these properties near excavation, Proc. 3rd Congr. Int. Soc. Rock Mechs., Denver, 393 p., 1974.

Brennan, B.J., Pulse propagation in media with frequency dependent Q, Geophys. Res. Lett., 7, 211-213, 1980.

Brennan, B.J., Linear viscoelastic behaviour in rocks. This volume, 1981.

Brennan, B.J., and F.D. Stacey, Frequency dependence of elesticity of rock: test of seismic velocity dispersion, Nature, 268, 220-222, 1977.

Gladwin, M.T., Precise in situ measurements of acoustic velocity in rock, Rivista Italiana di Geofisica, 22(516), 283-286, 1973.

Gladwin, M.T. and F.D. Stacey, Anelastic degradation of acoustic pulses in rock. Physics of the Earth and Planetary Interiors, 8, 332-336, 1974.

Grujic, N., Ultrasonic testing of foundation rock. Proc. 3rd Congr. Int. Soc. Rock Mechs., Denver, 404, 1974.

Just, G.D., and G.W. Walter, The engineering potential of seismic methods of assessing rock breakage characteristics. Aus. I.M.M. Melb. Branch Rock Breaking Symp., November, 1978.

Kjartansson, E., Constant Q - Wave propagation attenuation, J. Geophys. Res., 84 (B9), 4737-4748, 1979.

Meister, D., A new ultrasonic borehole meter for measuring geotechnical properties of intact rock, Proc. 3rd Congr. Int. Soc. Rock Mechs., Denver, 2, 410-417, 1974.

Murphy, V.J., Seismic velocity measurements for moduli determinations in tunnels, Proc. Nth.Am. Rapid Excavation and Tunnelling Conf. 1, 209, 1972.

Obert, L., R. Munson and C. Rich, Caving properties of climax ore body, Am. Inst. of Min. Met. and Pet. Eng. Trans., 260, 129-133, 1976.

Price, T.O., Demonstration of acoustical survey system, Washington Metro. Area Transit Authority, 1975.

Stacey, T.R., Seismic assessment of rock masses, Proc. Symp. Expl. Rock Eng., Johannesburg, 2, 113-117, 1976.

Volarovich, M.P., D.B. Balashov, I.S. Tomashevskaya, and V.A. Pavlogradsku, A study of the effect of uniaxial compression upon the velocity of elastic waves in rock samples under conditions of high hydrostatic pressure, Izv., Geophys. Ser., No. 8, 1198-1205, 1966.

Walsh, J.B., The effect of cracks on the compressibility of rock, J. Geophys. Res., 70, 381-389, 1965.

Winkler, K., A. Nur, and M.T. Gladwin, Friction and seismic attenuation in rocks, Nature, 277, 528-531, 1979.

FREQUENCY DEPENDENCE OF Q FOR ROCK STRESSED NEAR THE BREAKING POINT

Gary J. Turner and Frank D. Stacey

Department of Physics, University of Queensland, Brisbane 4067, Australia

Abstract. Frequency dependence of Q for samples of dolerite and granite stressed in flexure near to breaking point has been measured at seismic frequencies by rotating cantilevered rods. Stress is due to the bending moment of the weight of a rod, and strain appears as a vertical deflection of the end of the rod. Damping causes a lag in strain relative to stress and is apparent in this experiment as a horizontal deflection of the end of the rod, the deflection being opposite for opposite directions of rotation. Deflections arising from the bearings, which are estimated from measurements using nearly inflexible metal tubes as standard specimens, are appreciable and impose the principal limitation to the accuracy of the method, which is suitable only for specimens under high strain. The specimens were stressed near to the breaking point, and they usually broke after being rotated for an hour or so, although maximum strain amplitudes were only about 10^{-4}, rock being weak in flexure. Measured Q's were 125 for granite and 50 for dolerite, substantially lower than the values measured by different methods at low strain amplitudes for the same rocks, and were found to be independent of frequency over the range 0.007 to 0.6 Hz. It appears that an experiment such as the present one, in which the strain cycle involves tension near to the breaking point of the specimen, emphasises the role of grain boundaries, micro-cracks and pores, and that these give amplitude-dependent (non-linear) but frequency-independent damping. Frequency independence of the linear damping that occurs at very low strain amplitudes (and gives higher Q's) cannot be inferred from measurements at high strain amplitudes. The decrease in Q with the opening of cracks or pores may be relevant to earthquake prediction.

1. Introduction

The circumstances in which the anelastic Q of rock is observed to be independent of frequency have become important in determining the Q of the Earth. Evidence that Q of laboratory specimens is essentially constant over a very wide frequency range is well documented [e.g., Knopoff, 1964], although individual mechanisms of anelastic damping are mostly frequency dependent [Jackson and Anderson, 1970], and recent geophysical observations are increasingly interpreted in terms of frequency dependence [Anderson and Minster, 1979]. It is possible that the frequency independence is an artifact of the manner in which most laboratory measurements are made. Since it is now established that damping in laboratory samples at laboratory temperature is linear for strain amplitudes smaller than about 10^{-6} [Brennan and Stacey, 1977; Winkler et al., 1979], we can examine the possibility that frequency independence is a characteristic of the non-linear range and is not a property of the linear, lower amplitude range.

The experiment reported here is concerned with a novel method of seeking frequency dependence in the non-linear range. Rods of rock were stressed in flexure near to their breaking points, so that damping mechanisms due to friction at grain boundaries and pores would be enhanced by the opening (and closing) of cracks and pores to maximum extent. It is believed that such friction dominates high strain amplitude non-linear damping in rocks [Winkler et al., 1979].

2. The Rotating Cantilever Method of Measuring Q

Since the last century, rotation of a loaded cantilever has been a standard technique in fatigue studies on engineering materials. Samples are subjected to cyclic flexure simply by being rotated at any desired speed. It is not clear from the literature how early the horizontal deflections of the ends of the rotating samples were first recognised, but Mason [1923] developed the theory of the phenomenon, and Kimball and Lovell [1927] reported what were probably the first systematic measurements of anelastic Q by this method.

The theory is extremely simple. In response to the gravitational load of the sample itself and any added mass, the end of the sample is depressed, as represented by the distance h in Fig. 1. Then by virtue of rotation it is also deflected to A or B, depending upon the sense of rotation. This horizontal deflection is most easily measured by noting the relative displacement, 2δ, of the rod

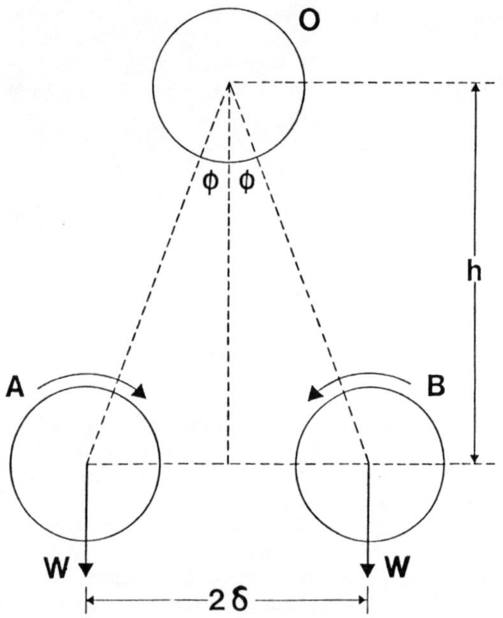

Fig. 1. End view of rotating cantilevered rod. O represents the position which it would have if it were not bent by its own gravitational weight and A, B are the alternative positions for opposite directions of rotation.

for opposite directions of rotation. It arises from the phase lag ϕ of the elastic flexure of the rod relative to the bending moment, where

$$\tan \phi = \frac{\delta}{h} = Q^{-1}$$

is identified directly in terms of the specific dissipation factor Q^{-1} of the sample under flexural strain at the frequency of rotation.

The method has been used to observe anelasticity in metals [Lazan, 1950] and more extensively for studies on polymers [e.g., Maxwell, 1956; McCammond, 1973] but not previously on rocks. Not all authors have been aware of a difficulty that arises when very small deflections are measured. Deflections also arise from the bearings in the lathes or similar machines that are used to hold the specimens. We have measured the bearing effect in a variety of machines with reversible rotations and find it to be measurable in all cases. The effect is typically 10 times greater for hydrodynamic bearings ($\sim$ 5 μm) than for rolling element bearings ($\sim$ 0.5 μm) or the one air bearing that we examined. Each bearing type also shows a characteristic frequency dependence of the effect (different for each type), so that to correct measured anelastic deflections, measurements of bearing deflection are required at each frequency.

3. Q Measurements on Dolerite and Granite

To ensure that the bearing effect was no more than a minor correction, we found it necessary to use specimens much longer than had originally been intended. Specimens 1.4 m long were made up by gluing together 20-cm lengths of 15-mm diameter granite and dolerite cores. The ends of each length were polished flat and bonded with an anaerobic glue. As a check on the effect of the glue, measurements with a similarly bonded length of steel rod were compared with those for a complete length of the same steel. Allowing for the greater bending moment of the steel, the horizontal deflections attributed to gluing of the rock specimens were estimated to be about 0.04 mm (independent of frequency), which is less than 10% of the deflections due to rock anelasticity and smaller also than the bearing effects.

Vertical deflections of the rods were not directly measurable but were calculated from their masses in terms of deflections measured for known forces and applied vertically upward to the ends of the rods by weights hung over light pulleys. Values of Young's modulus determined in this way for the samples used were 6.5×10^{10} Pa (0.65 Mbar) for dolerite and 4.4×10^{10} Pa (0.44 Mbar) for granite. Mean strain amplitudes (averaged along the specimens) were 0.9×10^{-4} (dolerite) and 1.3×10^{-4} (granite).

Results of Q measurements are represented graphically in Fig.2, in which frequency independence is apparent. Linear regressions to the data give gradients, with formal uncertainties, of $d \ln Q/d \ln f = (0.017 \pm 0.013)$ for granite and (0.003 ± 0.008) for dolerite.

4. Conclusions

Although the rotating cantilever method of measuring Q is limited to observations at high

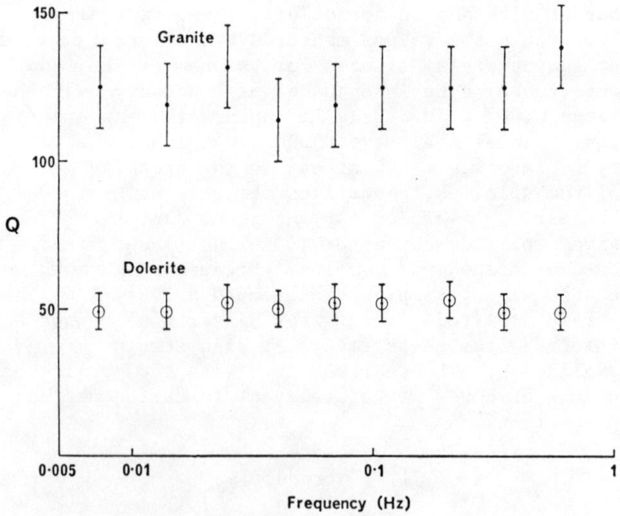

Fig. 2. Q vs. frequency for samples of granite and dolerite in flexure. The strain amplitude varies both radially and along the specimen being greatest at the surface and at the point of support. The average strain amplitudes at the surface are 1.3×10^{-4} for the granite and 0.9×10^{-4} for the dolerite.

strain amplitudes, it does provide a simple method of measuring the variation of Q over a wide frequency range in the seismic band and under identical cycled loads at all frequencies. The frequency dependence of Q is measured with greater certainty than the absolute value. No frequency dependence is apparent in our data. Assuming that a power law dependence is sought:

$$Q \propto f^n.$$

Then $n = d\,\ell n\,Q/d\,\ell n\,f$ is less than 0.02 in the dolerite and granite samples that we have examined.

We suggest that frequency independence is a characteristic of high strain amplitude measurements in which anelastic damping in rocks is dominated by cracks, pores, and grain boundaries. The fact that the material of rods cycled in flexure is alternately extended and compressed gives maximum opportunity for opening and closing of cracks and is probably responsible for the low Q values obtained. The granite sample was taken from the same block as one for which Brennan and Stacey [1977] reported a Q of 320 under cycled torsion at strain amplitudes of 10^{-6} or less. Thus the frequency dependence of Q under seismological conditions of low strain amplitudes cannot be inferred from measurements at high strain amplitude (even without doubts arising from temperature and pressure effects).

Although the specimens were stressed near to fracture and generally broke after no more than a few hours of testing, no time dependence of either elastic modulus (vertical deflection) or anelasticity (horizontal deflection) was noticed. Thus, apart from the fracture itself, we did not observe any irreversible evolution of the tested specimens. Nevertheless we identify the strong decrease in Q with the amplitude of a cycled strain in the range 10^{-5} to 10^{-4}, with effects which we may refer to collectively as "incipient cracks", and it is of interest to know whether decreasing Q is a less ambiguous indicator of dilatancy effects in seismic zones than are velocity variations. To test this question we need to measure Q of rocks under small cycled strains but high ambient (steady) strains.

References

Anderson, D.L., and J.B. Minster, The frequency dependence of Q in the Earth and implications for mantle rheology and Chandler wobble, Geophys.J., R. Astron. Soc., 431-440, 1979.

Brennan, B.J., and F.D. Stacey, Frequency dependence of elasticity of rock - test of seismic velocity dispersion, Nature, 268, 220-222, 1977.

Jackson, D.D., and D.L. Anderson, Physical mechanisms of seismic wave attenuation, Revs. Geophys. Space Phys., 8, 1-63, 1970.

Kimball, A.L., and D.E. Lovell, Internal friction in solids, Phys. Rev., 30, 938-949, 1927.

Knopoff, L., Q, Revs. Geophys., 2, 625-660, 1964.

Lazan, B.J., A study with new equipment of the effects of fatigue stress on the damping capacity and elasticity of mild steel, Trans. Am. Soc. Metal., 42, 499-549, 1950.

Mason, W., The mechanics of the Wohler rotating bar fatigue tests, Engineering, 15, 698-699, 1923.

Maxwell, B., An investigation of the dynamic mechanical properties of polymethyl methacrylate, J. Polymer Sci., 20, 551-556, 1956.

McCammond, D., Determination of dynamic compliance from creep data for polymethyl methacrylate, Plastics and Polymers, 41, 207-210, 1973.

Winkler, K., A. Nur and M.T. Gladwin, Friction and seismic attenuation in rocks, Nature, 277, 528-531, 1979.

ATTENUATION MECHANISMS AND ANELASTICITY IN THE UPPER MANTLE

Yves Gueguen*, Jacques Woirgard**, Michel Darot*

(*) Laboratoire de Tectonophysique, Université de Nantes, 44072 Nantes, FRANCE.

(**)Laboratoire de Physique des Matériaux, ENSMA, rue Guillaume VII, 86034 Poitiers Cedex, FRANCE.

Abstract. Anelastic attenuation of seismic waves in the earth's upper mantle is discussed with reference to laboratory data. It is likely that in a 3000km deep mantle with variable temperature, pressure, and mineralogy, several distinct mechanisms are active. Consequently only the upper mantle where olivine is the dominant phase is considered. The most appropriate experimental data are isothermal, high temperature, internal friction data. Internal friction experiments performed on a number of metals together with a few experiments performed on forsterite and peridotite emphasize that, in the solid state, dislocation relaxation mechanisms are most important. Such experiments question also the assumption of linearity. Recent work on dislocations in naturally and experimentally deformed olivine point to the fact that kink and jog nucleation are both difficult, even at high temperature, in that mineral. From these data, a dislocation model is suggested to interpret the seismic results for the upper mantle. In the model, anelastic and rheological properties of the upper mantle are connected through dislocation processes in olivine.

Introduction

Anelastic attenuation of seismic waves in the earth's mantle is high enough that it produces dispersion of the order of 1 % (Anderson et al., 1977) which has to be taken into consideration when velocity profiles are considered (Liu et al., 1976 ; Kanamori and Anderson, 1977). Although attenuation profiles have not yet the resolution of velocity profiles, models constructed from free oscillation and body waves show that attenuation is high both at the top and the bottom of the mantle (Anderson and Hart, 1978). In these regions, attenuation is usually assumed to be frequency independent although this is not clearly demonstrated. Understanding of the upper mantle anelasticity is of great importance since (1) there is a peak of attenuation around 100-200 km and (2) there may be some connection between that peak and the asthenosphere.

Physical understanding of these data is at its beginning. Phenomenological treatment of anelasticity is well known in the seismological literature. Since the review of Jackson and Anderson (1970), a few microscopical models have also been developed (Gueguen and Mercier, 1973 ; Anderson and Hart, 1978 ; Shaw, 1978). Investigation of anelasticity in solids has been under way for several decades in solid state physics laboratories. The results point to the fact that standard phenomenological and microscopical theories should be used with great caution. Two examples are significant : (1) when the hypothesis of linear behaviour is checked against experiments, it is disproved in a number of cases (see below I) ; (2) when the so-called "grain-boundary" peaks are investigated they are also found in single crystals (see below II).

"Relevant" experiments for anelastic attenuation are difficult to perform since they consist in measuring the factor Q^{-1} as a function of frequency, in the seismic band, at high temperatures and pressures, on appropriate minerals. A complete realistic interpretation of attenuation mechanisms in a 3000 km mantle with variable temperature, pressure and composition would require an impressive number of very difficult experiments. It is physically likely that in the different main regions of the mantle, different microscopic mechanisms dominate, depending on P, T, and the intrinsic natures of the minerals present at those depths. Consequently, and as a first step, we will restrict ourselves to the upper mantle (0-400), where olivine is believed to be the dominant mineral.

Internal friction in solids : phenomenological description and experimental methods

The different experimental methods used to measure internal friction correspond to different ranges of frequency, stress and temperature (Figure 1 and Table 1). Methods operating at high frequencies are not considered here.

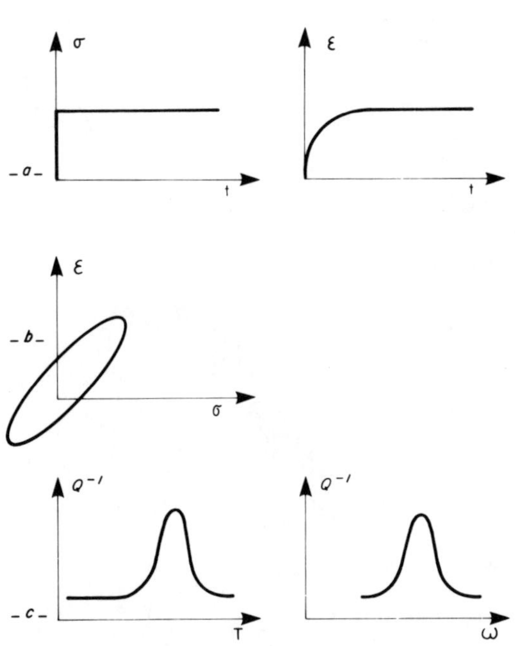

Fig. 1.
(a) Creep function and relaxation experiments : $Q^{-1}(\omega)$ can theoretically be derived from such data for a linear solid.
(b) Mechanical hysteresis loops : for a linear solid, the loops are theoretically elliptical.
(c) Classical pendulum and isothermal pendulum : Q^{-1} is directly measured, as of function of T or ω. No assumption is made on linearity.

The most direct method is the determination of (σ, ϵ), loops at low frequencies (10^{-3} - 10 HZ) and room temperature. The area of one loop is the absorbed energy per unit volume per cycle. If the behaviour is linear, the Boltzmann equation applies :

$$\epsilon(t) = \int_{-\infty}^{t} \dot{\sigma}(\tau) f(t-\tau) d\tau$$

where f is the "creep function" and the $\sigma - \epsilon$ curves are elliptical. Brennan and Stacey (1977) used such a method and obtained elliptical loops for basalt and granite at strain amplitudes of 10^{-6}. Peselnick et al (1979) obtained a similar result in granite at strain amplitudes of 10^{-5}. However these experiments, performed at room temperature, cannot be extrapolated to high temperatures since high temperature absorption phenomena are in general thermally activated.

Another method is to measure directly the creep function of the solid, that is its response to a step function stress. When the assumption of linearity is valid, the Boltzmann equation makes it possible to calculate $\epsilon(t)$ for any stress function $\sigma(t)$. Relaxation experiments have been performed by Berckemer et al (1979) on natural peridotite in order to measure the creep function f(t). However two important problems are raised by this kind of experiment. The first problem is that a considerable loss of accuracy is introduced by the Fourier transformation when $Q^{-1}(\omega)$ is calculated from f(t). This is emphasized by Figure 2 which shows that a complex spectrum $Q^{-1}(\omega)$ may be related to an apparently simple creep function. A second important problem results from the assumption of linearity itself. Measurements on metals and alloys have shown that their behavior is not linear in a number of cases (Woirgard et al, 1979). Figure 3 shows $Q^{-1}(\omega)$ spectra for an aluminum single crystal at high temperatures. Depending on the strain amplitude ($\epsilon \simeq 10^{-6}$ to 10^{-5}), different spectra are obtained which implies a non linear behavior. Moreover, in several cases, a frequency dependence

$$Q^{-1} = A\omega^{-n}$$

has been suggested and interpreted by a creep function

$$f(t) = a + bt^{n}$$

through the assumption of linearity. I has been shown recently (Gerland, 1979) from measurements

TABLE I. Conditions of Temperatures (T), frequency (ν), strain amplitude (ε), pressure (P) for the experiments of Figure 1 and the upper mantle.

	T	ν	ε	P
Upper mantle	> 0.5 Tm	10^{-4} - 10 s^{-1}	$\ll 10^{-7}$	20 - 150 kb
Relaxation creep	> 0.5 Tm	assumption of linearity	$> 10^{-6}$	0
Hysteresis loops	$\ll$ 0.5 Tm	10^{-4} - 10 s^{-1}	$> 10^{-7}$	0
Internal friction	> 0.5 Tm	10^{-4} - 10 s^{-1}	$> 10^{-7}$	0

of $Q^{-1}(\omega)$ over a much broader frequency range, that the behavior

$$Q^{-1} = A\omega^{-n}$$

may be representative of only a very small part of the spectrum (Figure 4).

The last group of experimental methods involves classical pendulums (flexure and torsion). Q^{-1} is obtained from the damping of the free oscillations of the system. Such methods are very sensitive since Q^{-1} values of 10^{-4} - 10^{-5} can be measured, although spurious damping may be problematical. The pendulums operate at a fixed frequency : if the absorption mechanisms are thermally activated with a relaxation time

$$\tau = \tau_o \exp\ (E/RT)$$

it is possible to simulate a variation of frequency by a variation of temperature. Obviously this is acceptable only if a temperature variation does not introduce any microstructural modification. This is not the case at high temperatures for most materials. For that reason Woirgard et al (1979) have proposed a new torsional pendulum (Figure 5) which allows us to make internal friction measurements as a function of frequency, at a constant temperature. The mechanical system is set into forced vibrations and the phase angle between stress and strain is measured from the time lag t :

$$\delta = \omega t, \qquad Q^{-1} = \tan \delta$$

This apparatus is indeed a "spectrometer" for elastic waves. In general, the observed spectrum $Q^{-1}(\omega)$, is resolved into several lines associated with distinct dissipative mechanisms. No

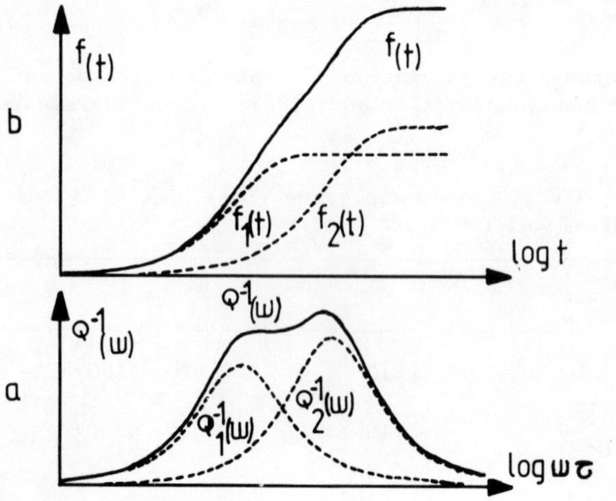

Fig. 2.
(a) Spectrum $Q^{-1}(\omega)$: 2 relaxation times τ_1 and τ_2 ($\tau_1/\tau_2 = 1/10$)
(b) Creep function
(from Nowick and Berry, 1972).

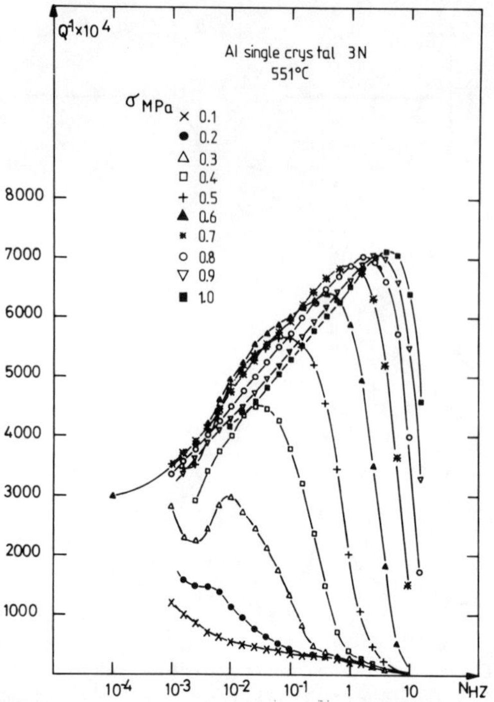

Fig. 3. Isothermal spectrum $Q^{-1}(\omega/2\pi)$ for aluminium single crystals at different strain amplitudes. For a linear solid, all the spectra should be identical.

linearity assumption is required, the frequency range covered is very large (10^{-6} - 10 Hz) and the apparatus can operate at high temperatures and low strain amplitudes (Table I). It appears to be the most suitable apparatus for geophysical purposes, although seismic waves operates at lower strain amplitudes and in high pressure conditions. Unfortunately these last conditions cannot be obtained presently by any method (Table I). Recent data obtained with our apparatus appear to throw some light on the potentially important attenuation mechanisms. A similar discussion published by Jackson and Anderson (1970) was based on data available at that time.

Internal friction data and microscopic mechanisms of attenuation in solids

We will restrict our discussion to the mechanisms that are potentially most important. Partial melting, fluid flow and phase transformation are not considered. They have been discussed recently by Mavko et al (1979). The data reported for these processes are not generally obtained at high temperatures and low frequencies. However, it is well known that very high attenuations can be associated with phase changes. Data have recently been reported for polymers or amorphous selenium (Gerland, 1979).

Absorption and attenuation appear when thermo-

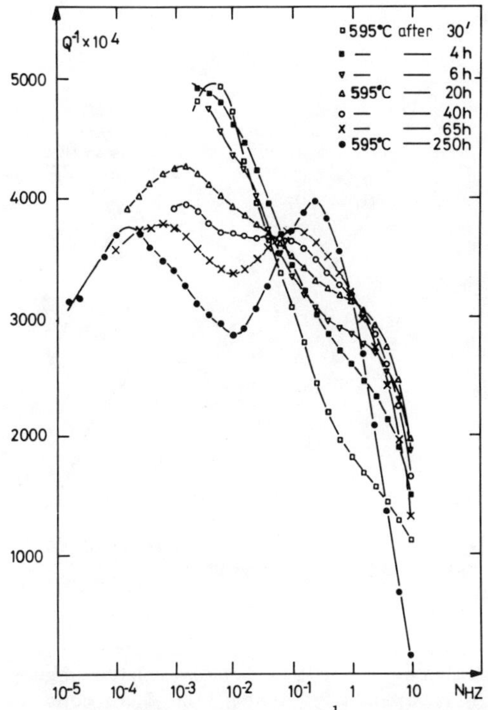

Fig. 4. Isothermal spectrum $Q^{-1}(\omega/2\pi)$ for aluminium single crystals at low strain amplitude (8.10^{-6}). Classical pendulums operating at constant frequencies allow only measurements between 1 - 10 Hz. In this region $Q^{-1}(\omega) \simeq \omega^{-n}$.

dynamically irreversible processes are present. Any departure from a perfect elastic behaviour is an example of such processes. It should be emphasized that various degrees of irreversibility can be found, related to the scale of the elementary processes involved. One extreme situation of very strong irreversibility is associated with "catastrophic" processes (for instance the break-away of a dislocation from pinning points). This type of process is strongly non linear. At the opposite, some situations are associated with a progressive, almost reversible, evolution of the material under the applied forces, in which case the processes may be linear. As an example one can mention diffusive processes.

Point defect relaxation

At least three types of point defects have been observed to produce absorption by relaxation : interstitial, subtitutional defects and vacancies. These processes are thermally activated. De Batist (1972) has discussed in detail the results obtained in different materials. The activation energies are related to diffusion of the interstitial or subtitutional species. In most cases they are low so that the damping is observed at low temperatures or high frequencies (Figure 6), as in the case of Fe-C or halides discussed by Shaw (1978). The intensity of the relaxation is a function of the defect concentration and is generally low. However Zener relaxation can produce high absorption in very concentrated solid solutions.

Vacancies can lead to attenuation through Nabarro creep. In that case vacancies flow in response to the chemical potential gradient set up by the deviatoric stress (Escaig, 1962). Diffusion is so slow in olivine(Poumellec et al, 1981) that such a process is unlikely in that mineral.

Viscous grain boundaries

Most of the effects observed above 0.5 Tm (Tm is the melting point) have, for a long time, been attributed to the viscosity of grain boundaries (Kê, 1947). Recent experiments on Cu, Al, Ag, Ni have demonstrated that these effects are indeed the result of dislocation motion inside the crystals (Woirgard, 1979 ; Rivière and Woirgard, 1977; Woirgard and De Fouquet, 1977). Figure 7 shows the result obtained on pure Cu. The "grain boundary" peak is also present in the single crystal.

Dislocation mechanisms

A large number of relaxation peaks result from the presence of dislocations (De Batist, 1972 ; Nowick and Berry, 1972). Three types of effects are recognized :

(1) Interaction between dislocations and point

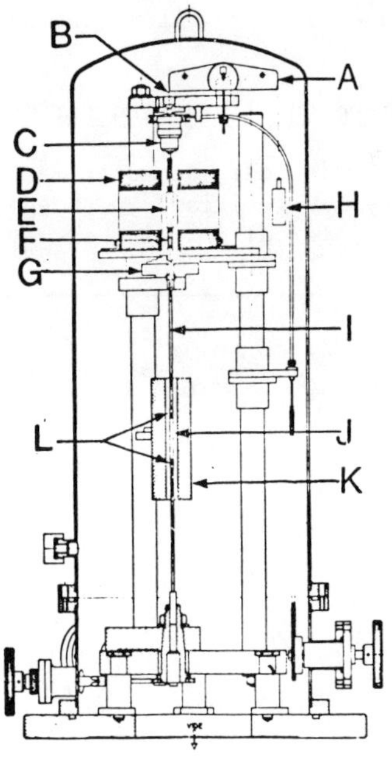

Fig. 5. Torsion pendulum (Woirgard et al., 1977) : A : Scales beam, B : suspension thread, C : mandrel, D : Helmoltz coils, E : permanent magnet, F : spherical mirror, G : mandrel, H : counter weight, I : extension rod, J : specimen, K : furnace, L : molybdenum grips.

defects produces low or intermediate temperature peaks. Activation energies are related to point defect diffusion.

(2) Dislocation jumps between two Peierls valleys (Bordoni peak) produce absorption at a temperature dependent on the double kink formation energy.

(3) Dislocation cross slip and dislocation climb by pipe or volume diffusion results in intermediate to very high temperature peaks (Woirgard, 1976, 1979). This includes the so called "grain boundary peaks".

A fourth effect, dislocation resonance, is not considered since it produces absorption only at very high frequencies (KHz-MHz range). It seems now that dislocation effects can explain the totality of the peaks observed at high temperatures (Woirgard, 1976 ; Woirgard and De Fouquet, 1977). The effects are not always linear as shown by Figure 3, even at low strain amplitudes. Very high attenuation are obtained in several cases (Figures 3 - 4).

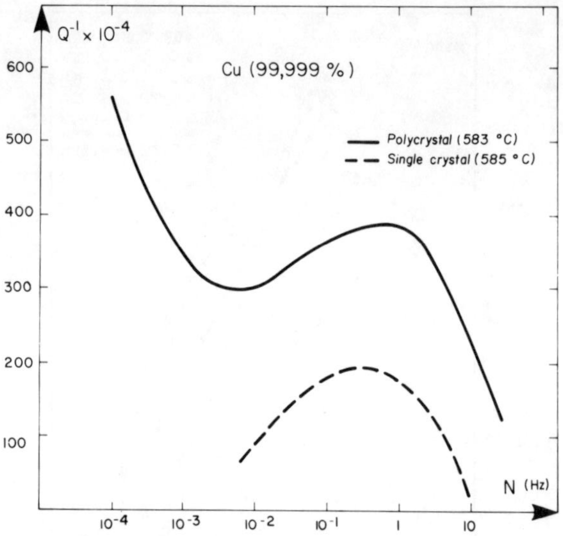

Fig. 7. "Grain boundary" peak observed in Cu single crystals and polycrystal.

Thermoelastic losses

This last mechanism does not involve any defect and is not thermally activated. It results from the heat flow which takes place between grain oriented differently or having different compositions and elasticities which experience different states of stress. This effect cannot be observed in metals at low frequencies due to their high thermal conductivities (De Batist, 1972). In ceramics, however, for a grain size of 1 mm the critical frequencies are within the seismic range ($\sim 10^{-1}$ Hz).

Results obtained on olivine and peridotite

Internal friction data on upper mantle materials are very few. Woirgard and Gueguen (1978) have measured internal friction in natural peridotite, synthetic polycrystalline forsterite and enstatite. Their measurements have been performed (1) between 20°C and 1100°C at constant frequencies (2-8 Hz) and (2) at low temperatures over a large frequency range (Figure 8-9). This investigation has shown that highly deformed peridotite exhibits a peak at 930°C (5 Hz). The peak is considerably reduced by annealing. Moreover, undeformed polycrystalline forsterite exhibits a background attenuation which slowly increases with temperature. The 930°C peak at 5 Hz could correspond to the increase in $Q^{-1}(\omega)$ at lower frequencies and lower temperature (Figure 9).

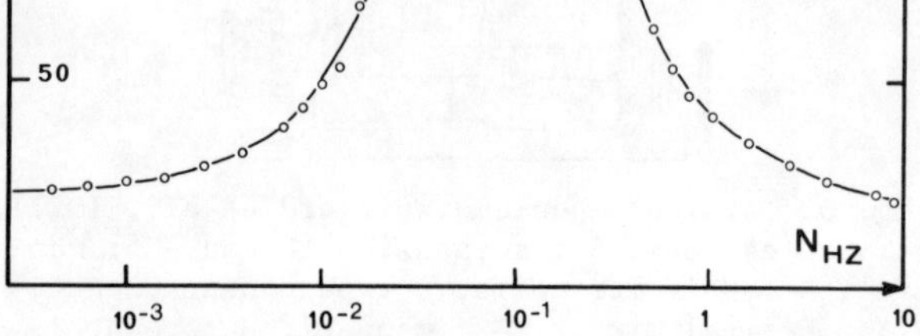

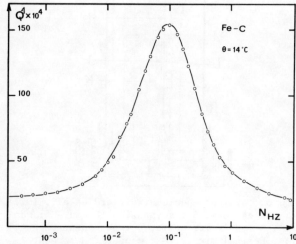

Fig. 6. Snoek effect observed in Fe. Spectrum $Q^{-1}(\omega/2\pi)$.Fe + 20 ppmC.

Although these experiments do not allow definitive conclusions, possible interpretations are that (1) the peak results from dislocation relaxation and (2) the background results from thermoelastic losses. Additional information is expected in the near future from experiments on single crystals deformed in laboratory and studied in a new torsional pendulum at high temperature and variable frequency. Such experiments are important but technically very difficult since small samples (< 2 cm) cannot be used in a pendulum. However single crystals are necessarily rather small samples.

Dislocation and anelastic properties of the upper mantle

Geophysical data

Several important results have been summarized by Dziewonski (1979) in a recent review paper : (1) between 1 hour and 10 seconds, Q of the earth appears to be frequency independent but between 10 and 1 seconds seismological data show some frequency dependence ; (2) in the upper mantle

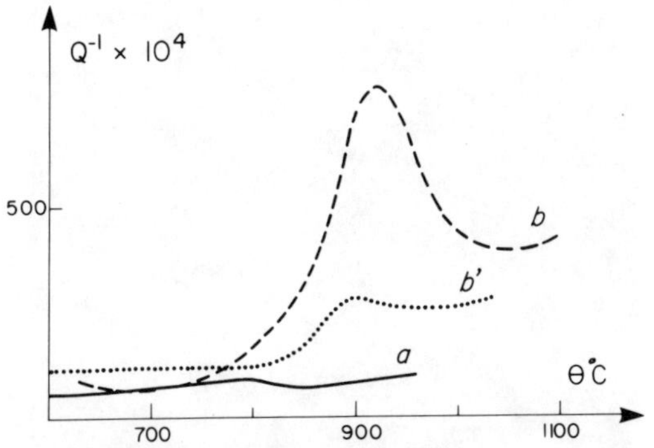

Fig. 8. Internal friction in peridotite and forsterite at constant frequencies :
(a) synthetic forsterite (1.8 Hz)
(b) naturally deformed peridotite (5 Hz)
(b') sample b after 5 hours annealing at 1100°C (flexural pendulum).
Strain amplitude = 10^{-6}.

low velocity zone, Q^{-1} values may be as high as 10 % ; (3) according to Anderson and Hart (1978), absorption is only due to shear in the mantle,but Dziewonski questioned this interpretation. The above results are of great importance for determining the physical mechanisms which produce absorption. A fourth important result can be added : the correlation between high attenuation and low velocity. Anderson and Hart (1976) have shown that after corrections for anelasticity, the upper mantle shear velocity profiles still show a pronounced drop around 170 km depth ($\Delta Vs/Vs \simeq -8$ %). Helmberger (1973) has modelled the low velocity zone for P waves and found a pronounced drop around 100 km ($\Delta Vp/_{Vp} \simeq -6$ %). According to this author, the sharpness of the boundaries of the zone are uncertain.

Unfortunately, detailed $Q^{-1}(z)$ profiles for different constant frequencies are still lacking. These data are fundamental since any frequency dependence can be averaged out by calculating $Q^{-1}(\omega)$ for a zone of several hundred of km thickness.

A model for interpretation of attenuation in the upper mantle

The geophysical data presented above and the physical data on dislocations in olivine (Gueguen, 1979 a and b) suggest a dislocation model for attenuation in the upper mantle. The correlation of high attenuation with low velocity, the observation of high attenuation and the fact that most of the absorption is due to shear, all point to a dislocation relaxation model. Absorption by dislocations cannot explain a bulk modulus effect. Moreover any dislocation relaxation model leads to both high attenuation and velocity drop

$$Q^{-1} \max = (\frac{\Delta V}{V}) \max = A$$

intensity of relaxation (Gueguen and Mercier, 1973). The question raised at this point is : what are the most important dislocation microstructures in olivine in the upper mantle ? For several reasons we suggest that the three microstructures described by Figure 10 apply to the upper mantle : (1) Figure 10 a and b are systematically observed in olivine and forsterite single crystals deformed at or above 1400°C between 50 and 800 bars (Durham et al, 1977 ; Darot et Gueguen, 1981), (2) they are also observed in naturally deformed polycristalline olivine, in particular in mantle peridotite xenoliths (Gueguen, 1979 a), (3) Figure 10 c is observed either in single crystals which have experienced inhomogeneous deformation or in naturally deformed polycrystals. The first two microstructures result from the fact that the dominant glide systems at high temperatures are (010) [100] and (001) [100] . The third one results from the necessary existence of local strain gradients in a deformed polycrystal.

As a consequence, two dislocation-relaxation mechanism corresponding to the microstructures of Fig. 10 a-b and Fig. 10 c) are considered to be realistic candidates to interpret Q^{-1} and $\Delta Vs/Vs$ data : (1) Bordoni peak, (2) dislocation climbin subboundaries. Relaxation by impurity dragging (Gueguen and Mercier, 1973) is less likely since this does not correspond to observations of dislocations : impurities seem to not modify the dislocation microstructures. Natural olivine and synthetic forsterite deformed experimentally in the same conditions have identical microstructures, although these crystals have obviously very different impurity contents. The dragging force which could be produced by impurities seem to be very small compared to that produced by lattice friction (Peierls force). Very high temperature peaks ($T \simeq 0.9$ Tm) (Woirgard, 1979) are not considered here since they do not correspond to the situation of the upper mantle. Although the geotherm is not well known, it is very likely that for olivine between 100 and 400 km depth, $T < 0.9$ Tm (Kennedy and Higgins, 1972) where Tm is the melting

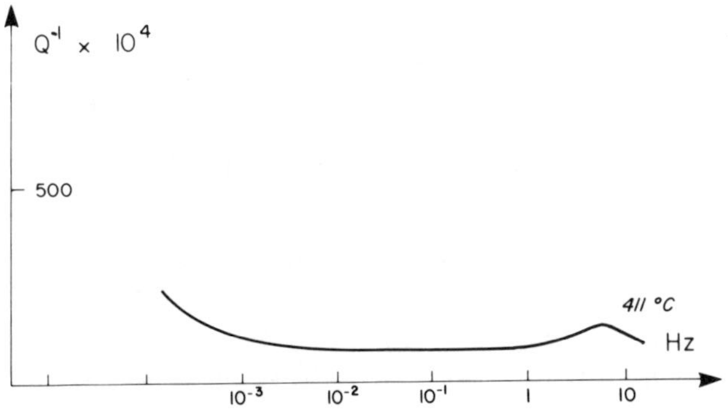

Fig. 9. Spectrum Q^{-1} (ω) in the sample of fig. 8b at 411°C (torsion pendulum).
Strain amplitude = 10^{-5}.

point of olivine (and not the solidus of the mantle).

The "Bordoni peak" is the result of the jump of dislocation segments AB = l between two Peierls valleys (Figure 11). In presence of an internal stress σ_i the work associated with a displacement h of AB is

$$\sigma_i blh$$

where b is the Burgers vector. If this quantity is equal to $2W_k$, double kink formation energy, the two configurations AB and AA'B'B (Figure 11) are equally probable. Consequently when an additional oscillating stress is applied, energy absorption results from oscillations of the dislocation between AB and AA'B'B. The absorption is maximum when the frequency is

$$\upsilon = \frac{2W_k}{kT} \quad \upsilon_D \exp\left(\frac{-2W_k}{kT}\right)$$

where υ_D is the Debye frequency, and k the Boltzmann constant (Hirth and Lothe, 1968). In metals, due to their low energy of kink formation ($2W_k \simeq 0.1$ eV), Bordoni peaks are observed at low temperatures and high frequencies. The existence of much higher Peierls forces in cova-

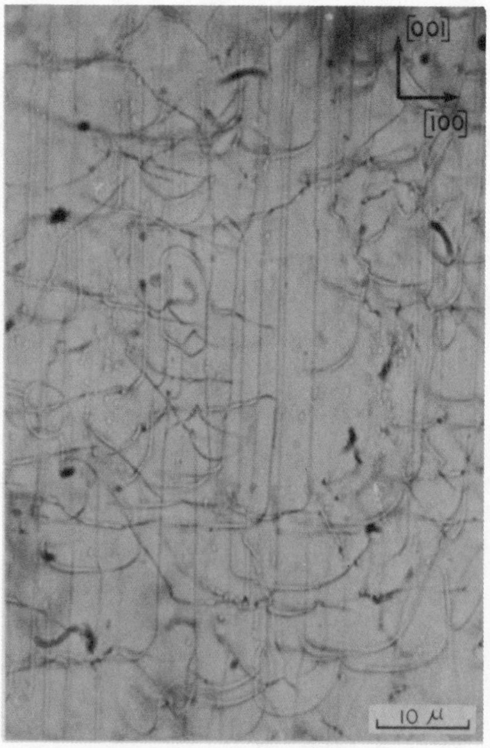

Fig. 10. The three typical dislocation microstructures in olivine and forsterite at high temperatures :
(a) (010) glide loops : straight, long, edge segments and short screws (forsterite single crystals deformed at 1600°C).

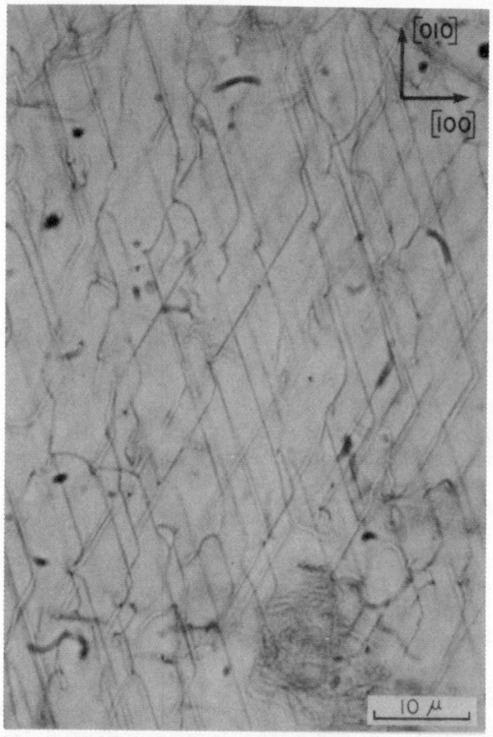

(b) (001) glide loops : straight mixed segments along <110> and short screws (forsterite single crystals deformed at 1500°C).

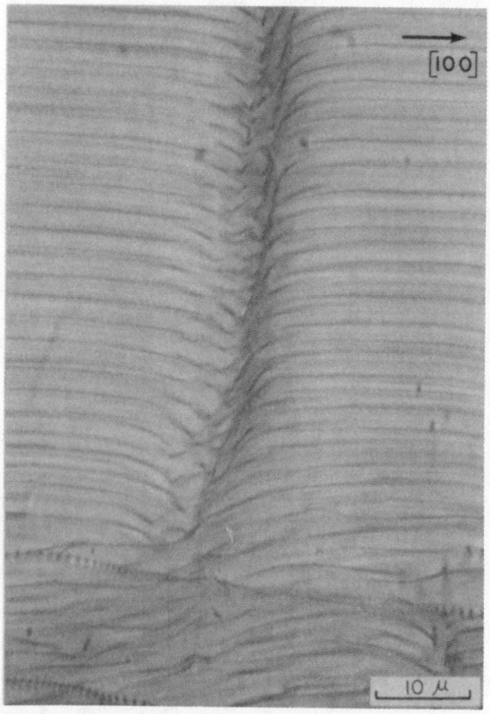

(c) "100 organization" : (100) tilt walls and [100] screws between two walls (olivine naturally deformed in a mantle xenolith).

lent crystals is well known (Friedel, 1965) so that we should expect higher values of $2W_k$. This is precisely what is suggested in olivine from the observations. The presence of numerous straight dislocation segments at high temperatures in olivine demonstrates that Peierls forces are high (figure 10 a and b). High values of $2W_k$ imply that the Bordoni peak will appear at high temperatures and low frequencies.

Dislocation climb could also produce high attenuation. In that case the absorption of energy results from jog migration under an external stress (Woirgard, 1976) (Figure 12). Climb can take place either by bulk or pipe diffusion. The first is very slow in olivine (Poumellec et al, this volume) : at 1600°C the diffusion distance for Si is of the order of b for periods of 30 s. Consequently an absorption effect at low frequencies could be observed only at very high temperatures ($>1600°C$). Moreover it would lead to a very small value for A, intensity of relaxation. No data on pipe diffusion are available. Free dislocations are very straight at high temperatures (Figure 10 a-b), suggesting that pipe diffusion is not active. However bound dislocations (tilt walls) observed in polycrystals (Fig. 10 c) are more irregular. Pipe diffusion could be active in that case and result in high internal friction at intermediate temperatures ($T \simeq 0.5$ Tm). This mechanism could provide a satisfactory explanation for the data reported on Figure 8 : the dislocation microstructure of the deformed peridotite presents a very large number of (100) tilt walls (average spacing 5 μm). The Q^{-1} peak observed at $T \simeq \frac{Tm}{2}$ is decreased by annealing although no modification of the free dislocation microstructure is visible. The relaxation time can be calculated from Woirgard (1976) :

$$\tau = \frac{kT}{D'} \frac{l^4}{\mu b^4 d}$$

where D' = pipe diffusion coefficient
μ = shear modulus
l = average free length of dislocation
d = average spacing between jogs.

Conclusions

Many microscopic mechanisms can contribute to attenuation of seismic waves in the upper mantle. Mechanisms involving several phases, in particular partial melting, have not been considered, although this last one is frequently quoted for explaining plasticity and anelasticity of the upper mantle. However it is now well known that high temperature olivine plasticity is by itself a very satisfactory explanation for flow in the mantle and partial melting is not required (Nicolas and Poirier, 1976, Goetze, 1977). It should be stressed also that no relevant data are available on this kind of system. On the other hand, many experimental results are available on attenuation mechanisms in the solid state. Isothermal low frequency internal friction data on a number of materials, including peridotite, emphasize that dislocation relaxations result in high values of Q^{-1}. Physical observations on dislocations in olivine and forsterite have greatly progressed in recent years. They suggest two possible relaxation mechanisms in this mineral : one at high temperatures (Bordoni peak) and another at intermediate temperatures (dislocation climb in subboundaries). Experimental data on olivine single crystals and on $Q^{-1}(z)$ profiles in the mantle for various frequencies are required for a definite interpretation of anelasticity-plasticity of the upper mantle.

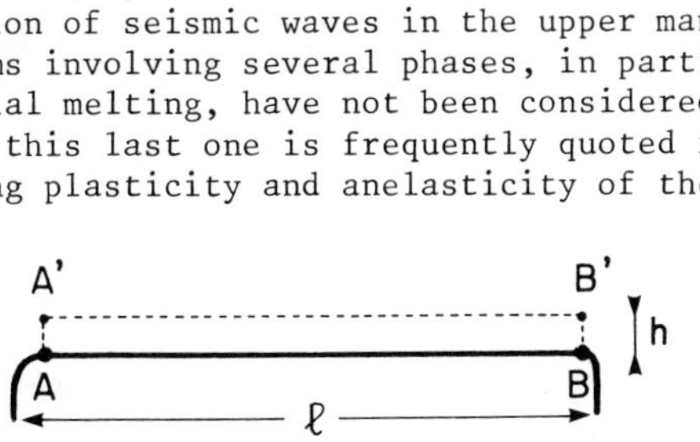

Fig. 11. Mechanism of absorption for the Bordoni peak : AB and AA'B'B are equivalent configurations.

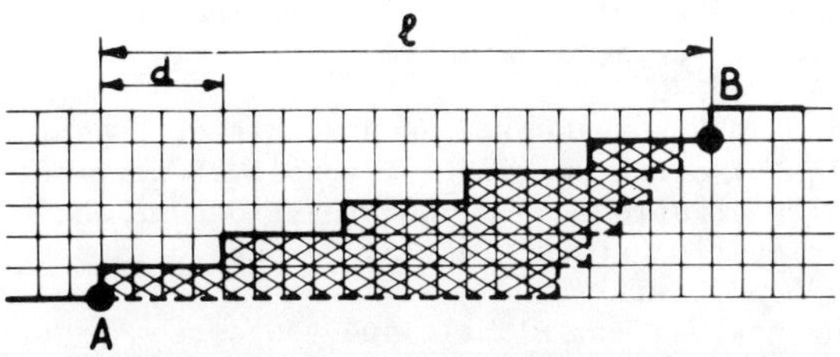

Fig. 12. Mechanism of aborption by jog migration : l is the dislocation free length and d the distance between jog.

References

Anderson, D.L., and R.S. Hart, Absorption and the low velocity zone, Nature, 263, 397-398, 1976.

Anderson, D.L., H. Kanamori, R.S. Hart, and H.P. Liu, The earth as a seismic absorption band, Science, 196, 1104-1106, 1977.

Anderson, D.L., and R.S. Hart, Attenuation models of the Earth, Phys. Earth Planet. Int., 16, 289-306, 1978.

Berckemer, H., F. Auer, and J. Drisler, High temperature anelasticity and elasticity of mantle peridotite, Phys. Earth Planet. Int., 20 (1), 48-59, 1979.

Brennan, B.J., and F.D. Stacey, Frequency dependence of elasticity of rock test of seismic velocity dispersion, Nature, 208, 220-222, 1977.

Darot, M., and Y. Gueguen, High temperature creep of forsterite single crystals. J. Geophys. Res. (in press).

De Batist, R., Internal Friction of Structural Defects in Crystalline solids, Elsevier, 477 p., 1972.

Durham, W.B., and C. Goetze, Plastic flow of

oriented single crystals of olivine, 1. Mechanical data, J. Geophys. Res., 82, 5737-5753, 1977.
Durham, W.B., C. Goetze, and R. Blake, Plastic flow of oriented single crystals of olivine, 2. Observations and interpretations of the dislocation structures, J. Geophys. Res., 82, 5755-5770, 1977.
Dziewonski, A.M., Elastic and anelastic structure of the earth, Rev. Geophys. Space Phys., 17 (2), 303-312, 1979.
Escaig, B., Frottement intérieur de haute température et diffusion de lacunes entre les dislocations, Acta Met., 10, 829-834, 1962.
Friedel, J., Dislocations, Pergamon Press, Oxford, 491 p., 1964.
Gerland, M., Corrélation entre la structure et le comportement en frottement intérieur isotherme et en microfluage d'un aluminium monocristallin, Thèse de 3ème Cycle : Poitiers, 68 p., 1979.
Goetze, C., Bounds on the subsolidus attenuation for four rocks types at simultaneous high pressure and temperature, Tectonophysics, 42, T1-T5, 1977.
Gueguen, Y., and J.M. Mercier, High attenuation and the low velocity zone, Phys. Earth Planet. Int., 7, 39-46, 1973.
Gueguen, Y., Dislocations in naturally deformed terrestrial olivine : classification, interpretations, applications, Bull. Soc. Fr. Mineral. Cristall., 102, 178-183, 1979 a.
Gueguen, Y., High temperature olivine creep : evidence for control by edge dislocations, Geophys. Res. Letters, 6, 357-360, 1979 b.
Helmberger, D.V., On the structure of the low velocity zone, Geophys. J.R. Astron. Soc.,34, 251-263, 1973.
Hirth, J.P., and J., Lothe, Theory of dislocations, Mc Graw-Hill, N.Y. 780 p., 1968.
Jackson, D.D., and D.L., Anderson, Physical mechanisms of seismic wave attenuation, Rev. Geophys. Space Phys., 8, 1-63, 1970.
Kanamori, A., and D.L. Anderson, Importance of physical dispersion in surfaces waves and free oscillations problems, Rev. Geophys. Sp. Phys., 15, 105-112, 1977.
Ke, T.S., A grain boundary model and the mechanism of viscous intercrystalline slip, J. Appl. Phys. 20, 279, 1949.
Kennedy, G.C., and G.H. Higgins, Melting temperatures in the earth's mantle, Tectonophysics, 13, 221-233, 1972.
Kohlstedt, D.C., C. Goetze, W.B. Durham and J.B. Vandersande, A new technique for decorating dislocations in olivine, Science, 191, 1045-1046. 1976.
Liu, H.P., D.L. Anderson and H. Kanamori, Velocity dispersion due to anelasticity ; implications for seismology and mantle compositions, Geophys. J.R. Astron. Soc., 47, 41-58, 1976.
Mavko, G., Kjartansson, E., Winkler, K., Seismic wave attenuation in rocks, Rev. Geophys. Sp. Phys. 17, 1155-1164, 1979.
Nicolas, A., and J.P. Poirier, Crystalline plasticity and solid state flow in metamorphic rocks, Wiley, 1976.
Nowick, AS., and B.S. Berry, Anelastic relaxation in crystalline solids, Academic Press. N.Y. 1972.
Peselnick, L., H.P. Liu, and K.R. Harper, Observation of details of hysteresis loops in westerly granite, Geophys. Res. Lett., submitted, 1979.
Poumellec, M., O. Jaoul, and C. Froidevaux, Diffusion and creep in forsterite, this volume.
Rivière, A., and J. Woirgard, High temperature internal friction peaks in polycristalline and monocristalline pure silver, VI th International Conference on internal friction. University of Tokyo Press. 749. 1977.
Shaw, G.H., Interpretation of the low velocity zone in terms of the presence of thermally activated point defects, Geophys. Res. Letters, 5, 629-632. 1978.
Woirgard, J., Modèle pour les pics de frottement interne observés à haute température sur les monocristaux, Phil. Mag. 33, 623. 1976.
Woirgard, J., Y. Sarrazin, and H. Chaumet, An apparatus for the measurement of internal friction as a function of frequency between 10^{-6} Hz and 10 Hz, Rev. of Sc. Instr. 48, 1322, 1977.
Woirgard, J., and J. De Fouquet, High temperature internal friction measured as a function of frequency between 10^{-5} Hz and 10 Hz on high purity metals, VIth Int. Conference on Internal friction. University of Tokyo Press, 671, 1977.
Woirgard, J., and Y. Gueguen, Elastic modulus and internal friction in enstatite, forsterite and peridotite at seismic frequencies and high temperatures, Phys. Earth Planet. Int.17, 140-146, 1978.
Woirgard, J., M. Gerland, and A. Rivière, To be published in "Proceedings of the IIIrd E.C.I.F.U.A.S. - Manchester - 1979.

SILICON DIFFUSION IN FORSTERITE: A NEW CONSTRAINT FOR UNDERSTANDING MANTLE DEFORMATION

Olivier Jaoul, Michèle Poumellec, Claude Froidevaux and Andrée Havette

Laboratoire de Géophysique et Géodynamique interne, Université Paris Sud
Orsay 91405, France

Abstract. The extrapolation of laboratory data on creep of olivine to upper mantle conditions requires knowledge of the mechanism of dislocation motion. The most popular model is based on dislocation climb. In its simplest form it predicts the same temperature dependences for creep and for diffusion of the slowest atomic species. This work presents self-diffusion data for oxygen and silicon for different temperatures and oxygen fugacities, p_{O_2}. Oxygen diffusivity has an activation energy of 90 kilocal/mol and is not sensitive to p_{O_2}. It is five orders of magnitude lower than Mg diffusivity. Silicon diffusivity is by far the weakest, three orders of magnitude lower than for oxygen. All diffusion activation energies are definitely smaller than that of creep which amounts to 160 kilocal/mol. This discrepancy contradicts to above-stated climb mechanism for creep. On the other hand, creep, which is also p_{O_2} independent in forsterite, exhibits a $(p_{O_2})^{1/6}$ dependence in natural olivine. This leads to the conclusion that the point defect chemistry, and therefore atomic diffusion, take part in the deformation mechanism. As silicon has been found to be slower than oxygen, the guesses on the most probable pressure dependence of creep will also need some revision.

Introduction

Deformation processes in the Earth's upper mantle are at the origin of all large-scale tectonic events observed in geology. Their understanding relies on our knowledge of the creep properties of olivine and similar materials. Laboratory data for natural olivine (see, e.g., Durham and Goetze, 1977) and for forsterite single crystals (Durham et al., 1979) yield a deformation law believed to be applicable to upper mantle dynamics. The extrapolation of laboratory data to the Earth raises several questions, in particular that of the fundamental mechanism for creep in single crystals. What dislocation motion controls creep: glide or climb? What is the relationship between creep and atomic diffusion? This last point is a key problem for geodynamics. Its clarification will tell us whether the pressure dependence of mechanical processes can be derived from diffusion data. This property is essential for lithosphere-asthenosphere models (Froidevaux and Schubert, 1975).

Steady state creep in olivine has been interpreted in terms of dislocation climb (Weertman, 1970). This mechanism is controlled by atomic diffusion. It has been suggested that the largest ion, i.e. oxygen, was likely to be the slowest. Thus, one expected the activation energy for creep to be similar to the activation energy for oxygen self-diffusion (Stocker and Ashby, 1973; Weertman, 1975). Along this line of reasoning, but in the absence of experimental data, one expected the pressure dependence of creep to be given by an activation volume roughly equal to the volume of an oxygen ion. The compatibility with this conclusion of some creep experiments under pressure (Ross et al., 1978) does not prove all the above assumptions to be true.

Indeed this paper will show that silicon diffusion in forsterite single crystals is a much slower process than oxygen diffusion. Furthermore the activation energies for diffusion of both oxygen and silicon, not to mention magnesium, are much smaller than the activation energy of creep in the same material. These new data call for a revision of the simple ideas mentioned above concerning the relationship between creep and diffusion. Despite its small size, silicon could have been expected to diffuse very slowly: its strong covalent bonding within SiO_4 tetrahedra seems to be the explanation. These tetrahedra are known to be very stable in shape and size as observed by X-rays during thermal expansion of the crystal (Lager and Meagher, 1978).

The choice of forsterite Mg_2SiO_4 rather than iron-bearing natural olivine in the present study is related to several experimental considerations. The forsterite single crystals contain very few dislocations and are chemically pure. The dislocations have a preferred <100> orientation and are connected with the growth of the crystal. Their low density, observed by X-ray topography, a priori prevents the existence of efficient short circuits for diffusion. One expects therefore to measure diffusion in the bulk rather than through the cores of the dislocations. The absence of iron extends the range of oxygen-partial pressures in which the sample remain chemically stable. The complication of Fe^{2+} to Fe^{3+} transformations is avoided, but of course the difference in point defect chemistry should be kept in mind when the forsterite data are extrapolated to natural conditions. The experiments presented here were made on 18 samples all cut from the same single crystal. The creep data used for comparison, as well as the oxygen diffusion data, were obtained with the same material.

We first present the Si self-diffusion data and then describe some pertinent aspects of the experimental procedures. New oxygen self-diffusion (Jaoul et al., 1980; Reddy and Cooper, 1978) and creep data (Michaut, 1980) for forsterite will be described briefly before discussing the implications of our results.

Silicon Diffusion Results

The diffusion of ^{30}Si along three crystallographic directions has been experimentally determined. The diffusivity D^* for this isotope is related to self-diffusivity D of silicon by the relationship $D^* = fD$, where the correlation coefficient f is of the order of one. Its exact value depends upon the nature of the defects, vacancies or interstitials, involved in the diffusion process. The measured diffusivity has the form:

$$D^* = D^*_o \exp - \left[\frac{E}{RT}\right]$$

where T is the temperature and R the usual gas constant, the pre-exponential term $D_o = 1.5\ 10^{-6} cm^2/s$ and the activation energy $E = 90 \pm 10$ kilocal/mol. Figure 1 shows the measured values log D vs. 1/T. The above values of the parameters are derived from the straight line through the data in this plot. The temperature range is 1300 to 1700°C.

Several striking features appear in this first data set obtained for silicon diffusion in a silicate mineral. First the absolute value of the diffusivity is here about three orders of magnitude lower than that of oxygen in the same forsterite. A second feature seen in Figure 1 is the absence of crystalline anisotropy, within error bars, of the silicon diffusion. The possible variation of the diffusivity of silicon with oxygen-partial pressure surrounding the sample during annealing has also been investigated. Figure 2 shows that no systematic variation is detected when p_{O_2} is varied between atmospheric pressure and 10^{-10} atmospheres.

A great deal of experimental skill is involved

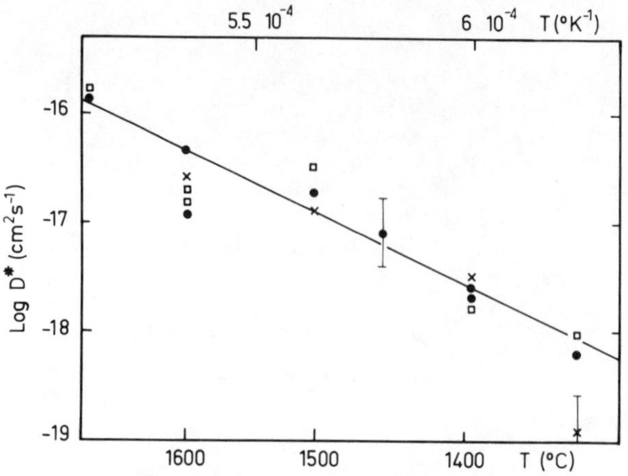

Fig. 1. Diffusion coefficient D versus inverse of temperature for ^{30}Si in forsterite. The slope of the straight line through the data points yields an activation energy of 90 kilocal/mol. Three different symbols are used for diffusion along different crystallographic orientations. Some error bars have been drawn. Where data points are clustered together, they have been omitted. Full circles are for [100] orientation, crosses for [010], squares for [001].

in this sort of measurement. The reader interested in the geodynamical implications may not want to read the next section concerning some important aspects of the methodology. For him one word is needed here about the amplitude of the errors bars in the above figures. These error bars are indeed very small considering the difficulty of the measurements. The problem was that the quantity to be determined turned out to be so small. The penetration length of ^{30}Si into the sample after an annealing time t is given by $2\sqrt{Dt}$. At a temperature of 1450°C a heat treatment of t = 1 month only yields a penetration of 700 Å. This represents a challenge to the experimentalist as to the crystalline quality of the surface of the samples and as to the appropriate method to measure the diffusion profile. Multiplying the above duration of the heat treatment by a factor of 10, which is difficult because of the lifetime of high temperature equipment, would only extend the penetration by $\sqrt{10}$ and does not change the nature of the experimental difficulty in measuring the profiles.

Experimental Procedures

The diffusion profiles of ^{30}Si into forsterite single crystals were determined by means of an ion probe (Slodzian, 1975): a beam of O_2^+ ions digs a crater into the sample surface and the extracted silicon ions are counted through a mass spectrometer. Natural abundances are 92.

27% for ^{28}Si, 4.68% for ^{29}Si, and 3.05% for ^{30}Si. Our source material is silica, SiO_2, enriched to 90% ^{30}Si. It cannot be deposited directly onto the forsterite surface because SiO_2 is soluble in forsterite (Pluschkell and Engell, 1968). We choose to synthesize forsterite by high temperature solid reaction of MgO with the enriched silica. After 48 hours at 1500°C, X-ray analysis showed that the chemical reaction was complete. An excess in MgO was always deliberately present in our experiments where a thin layer (100 Å) of enriched forsterite was deposited onto the sample by sputtering. MgO is not soluble in forsterite and this phase fixes the thermodynamical state of both the enriched layer and the bulk sample during the self-diffusion annealing procedure. The original surface of the sample has to be mechanically and chemically prepared so that good crystalline quality is observable in the first 30 Å (Jaoul et al., 1980).

The ion probe analysis records the number of ions of mass 28 and 30. It is checked that their sum is constant vs. depth of the crater. The latter is limited laterally by the meshes of an Al-grid laid onto the sample to prevent charge effects during the analysis. One hour of O_2^+ bombardment removes about 3000 Å of material. Figure 3 shows the profile of one crater, which is measured mechanically (Talystep) at the end of each analysis. It yields the exact rate of sputtering, usually with 10% accuracy or better.

The diffusion profiles corresponding to the imposed boundary condition should be of the form:

$$C(x) - C_\infty = \frac{M}{\sqrt{D^* t}} \exp\left[-\frac{x^2}{4D^* t}\right]$$

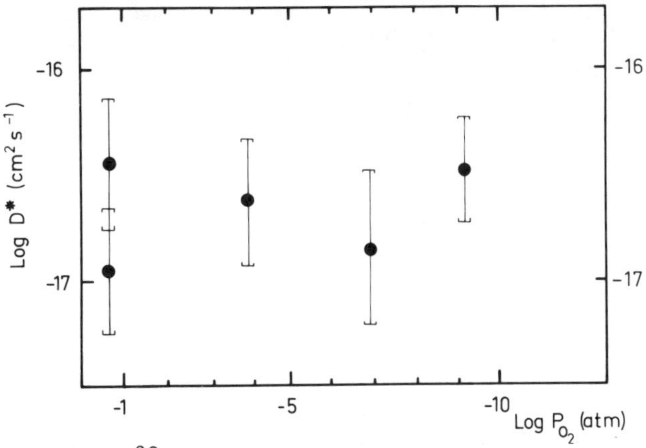

Fig. 2. ^{30}Si diffusion versus oxygen partial pressure in the atmosphere surrounding the sample during annealing at 1600°C. The diffusion takes place in the [100] direction. No dependence is detectable over a variation of p_{O_2} by nine orders of magnitude.

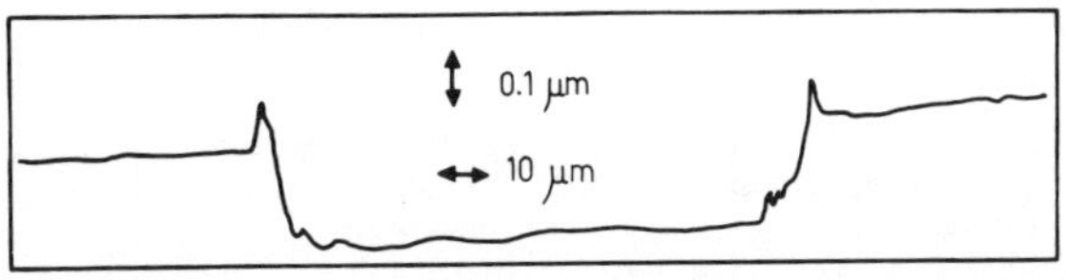

Fig. 3. Shape of a crater due to sputtering of the forsterite surface. The depth is about 0.2 µm and the surface about 120 (µm)^2. A thin mechanical point scans the sample surface to measure this profile. The two peaks next to the crater are what remains of an aluminium grid, which has been detached from the sample surface before this profile was measured. Only those ions extracted in the middle portion of this crater are analized, in the mass spectrometer (diaphragm of 30 µm diameter).

where M is the number of ^{30}Si atoms deposited on the sample per unit area, t the duration of the diffusion annealing, x the depth below the sample surface, and D* the diffusivity. The concentration C_∞ corresponds basically to the natual abundance of ^{30}Si, but its experimental value may differ slightly from it because of the presence of polyatomic complexes having the same mass as the collected silicon ions. Figure 4 gives an example of a diffusion profile C(x) as well as a representation of the same data in the form $\log(C - C_\infty)$ vs. x^2. The slope of the straight line, which yields the value of D* , does not depend critically upon the exact value of C_∞, but the data points depart from the straight line at large x^2 values if C_∞ has not been correctly adjusted.

Another significant correction derives from the fact that a step function, i.e., the initial ^{30}Si distribution before annealing, is not resolved as a sharp step by the ion probe analysis but rather as a smooth transition from the enriched value to C_∞ . The width of this transition, typically 200 Å, is found to be a small but sizable fraction of the measured diffusion profiles, so that a deconvolution of the latter is necessary. The physical origin of this broadening is related to both the fact that samples do not have a perfectly planar surface, and to a disturbance of the ^{30}Si distribution by the primary beam of O_2^+ ions. A dummy unannealed sample has thus to be compared with each diffused sample.

Comparison With Oxygen Diffusion And With Creep Data

It was mentioned above that oxygen diffuses about 10^3 times faster than silicon. Figure 5 shows some of the data (Jaoul et al., 1980). Here too, no p_{O_2} dependence is observed and one has $D_o = 10^{-4}$ cm^2/sec and E = 77 ± 10 kilocal/mol, a value close to the activation energy for silicon diffusion. It may be recalled that the cations have diffusivities several orders of mag-

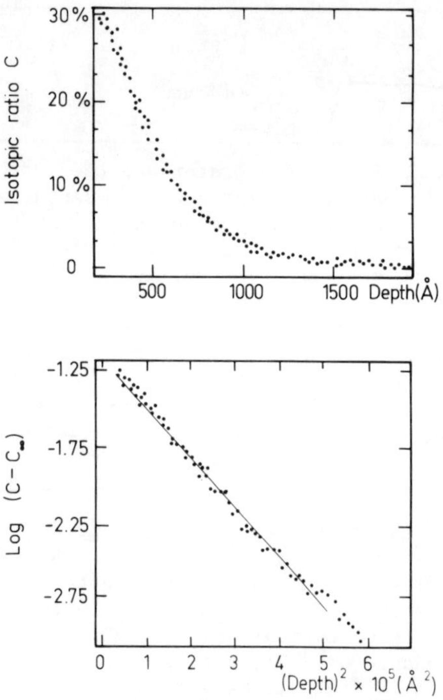

Fig. 4.(a) Measured isotopic ratio $^{30}Si/(^{30}Si + ^{28}Si)$ versus depth below the original sample surface. This sample was annealed in air (p_{O_2} = 0.21 atm.) at T = 1600°C for 1.61 10^6 sec. The initial portion of the profile is disturbed by transient electrostatic effects perturbing the O_2^+ primary beam. The ratio reaches the value C_∞ at great depth.
(b) Part of the same data is plotted here as $\log(C - C_\infty)$ versus depth squared. The straight line has a slope equal to $(4D^*t)^{-1}$ where t is the annealing time.

nitude greater than oxygen (Buening and Buseck, 1973). The temperature dependence of diffusion, particularly that of silicon, the slowest species, has now to be compared with the temperature dependence of creep.

At high temperatures and at moderate stresses of 100 to 1000 bars, natural olivine has a creep rate $\dot{\varepsilon}$ given by (Durham and Goetze, 1977).

$$\dot{\varepsilon} = A\sigma^n \exp\left[-\frac{Q}{RT}\right], \text{ with } n \simeq 3$$

Here RT means the same as in the diffusivity formula above, Q is the activation energy for creep and A an experimentally determined proportionality constant. For natural olivine Q is about 125 Kcal/mol. For forsterite the value is higher, 160 ± 7 Kcal/mol for creep experiments carried out between temperatures of 1500 to 1700°C (Durham et al., 1979). We have now found that the temperature variation of $\dot{\varepsilon}$, for a wider range of temperatures, extending as low as 1200°C, cannot be expressed by a single value of Q. Figure 6 shows indeed that $\log\dot{\varepsilon}$ vs 1/T does not follow a straight line: this experimental curve yields a slope corresponding to Q = 120 kilocal/mol at lower temperatures. A similar trend was known to exist in the data of olivine (Durham, 1975).

No detectable p_{O_2} dependence is found in the creep properties of forsterite, in contrast to the $(p_{O_2})^{1/6}$ dependence of $\dot{\varepsilon}$ in natural olivine (Jaoul et al., 1979). This last experimental result in iron bearing olivine is important as it strongly establishes that creep is sensitive to the point defect chemistry. Indeed the point defect concentrations ought to vary with p_{O_2} (Stocker, 1978). Hence there must be a relationship between creep and diffusion. In forsterite this could not be seen, as no measured property is p_{O_2} sensitive. In this mineral, in the absence of iron, the main defects are Mg vacancies and interstitials. This proposal is compatible with ionic conductivity data and with our self-diffusion results for oxygen and silicon.

We thus have to face the fact that the slower diffusing species, silicon, has an activation energy of 90 kilocal/mol in a range of temperatures at which the activation energy of creep is definitively much higher and varies between 160 and 120 kilocal/mol. In consequence the simple relationships connecting diffusion to dislocation climb, and the latter to high temperature deformation do not hold. From this negative conclusion, it also follows that no educated guess can be made about the pressure

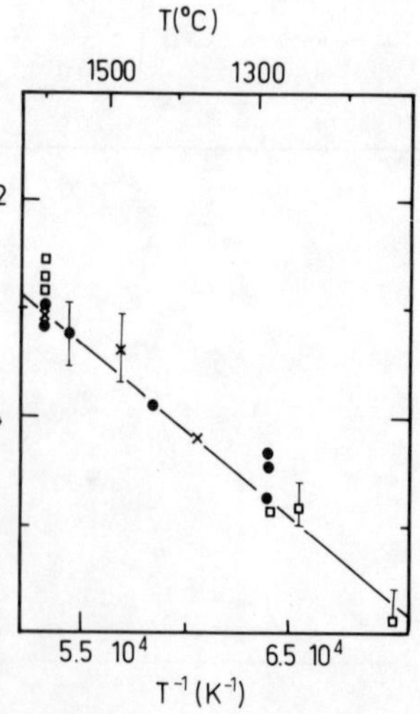

Fig. 5. ^{18}O diffusion coefficient versus inverse temperature for the same p_{O_2} = $10^{-9.2}$ atmosphere. Point data are given with the same symbols as in Figure 1. The straight line has a slope corresponding to an activation energy of 77 kilocal/mol. This value is not changed for other p_{O_2} conditions.

dependence of creep. Neither the atomic volume of oxygen can be retained to represent the activation volume of creep, nor should that of silicon be proposed at this stage.

Some Speculations On The Creep Mechanism

What is the possible relationship between silicon diffusion and creep in forsterite, and more generally in olivines? One first line of thought is to try to stick to the climb mechanism of edge dislocations. The discrepancy of the two activation energies found in our experiments could be handled by proposing that the energy of jog formation explains the difference (Gueguen, 1979). Indeed the emission or absorption of point defects along an edge dislocation occurs at places where the dislocation line makes steps, the so-called jogs. In metals the energy of jog formation is small and it seems somewhat unrealistic to invoke some 70 kilocal/mol for forsterite. Another suggestion has been made by Weertman (private communication) following our results on oxygen diffusion. It states that high temperature creep could be governed by diffusion through the bulk, but that the self-diffusion experiments measure diffusion through the dislocation cores. This argument implies that usually the activation energy for pipe diffusion is about half of that for bulk diffusion. It is not in agreement with the fact to our forsterite samples contain a very low density of dislocations (10^4 cm^{-2}). Further more these dislocations, which are linked with the growing process of the crystal, have a marked preferred orientation <100> . The average distance between dislocations is thus 100μm , a much larger value than our characteristic penetration lengths. Dominant pipe diffusion would lead to a strong anisotropy for D in favor of <100> . This is not observed.

So far our discussion has ignored the abundant information made available by using appropriate dislocation decoration techniques in deformed olivine (Kohlstedt, 1976) or forsterite (Jaoul et al., 1979). The observed patterns have hinted at possible dissociations of the dislocations. Two main features emerge (Durham, 1975; Gueguen, 1979; Michaut, 1980). (1) Conspicuous arrays of parallel straight <100> screw dislocations pinned in (100) tilt walls are seen. A proposal for a creep mechanism controlled by cross slip (Poirier and Vergobbi, 1978) is adapted to this situation. (2) More often the dominant dislocation pattern contains a large number of straight edge dislocations with Burgers vector [100] and glide plane (010). Here a possible climb could assist their mobility in the gliding plane but the activation energy of creep might be bigger than that of Si diffusion alone in this situation.

Conclusion

A new set of data has been added to our knowledge of the physical behaviour of olivine in

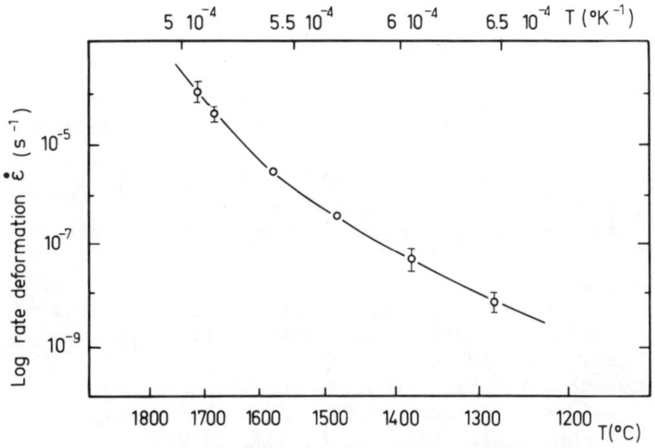

Fig. 6. Rate of deformation of a forsterite single crystal versus inverse temperature. The slope of the curve through the data points yields the activation energy of creep. It is seen to decrease with temperature.

conditions which can be compared with those in the Earth. Here the rates of atomic self-diffusion in forsterite have been experimentally determined. In particular the diffusivity of silicon is found to be much slower than that of oxygen. At 1600°C for example the comparison yields:
silicon: $5\ 10^{-17} cm^2/s$, oxygen: $10^{-13} cm^2/s$ and magnesium: $6\ 10^{-9}\ cm^2/s$. More generally the silicon results give:

$$D = 1.5\ 10^{-6} \exp\left[-\frac{90 \pm 10\ \text{Kcal/mol}}{RT}\right]$$

where the preexponential constant is in cm^2/s. The activation energy of self-diffusion is about equal to the energy of a Si-O bonding. Forsterite is very much like a hexagonal close packed structure of oxygen atoms with silicon atoms in tetrahedral positions and magnesium atoms in octahedral sites. The results thus suggest that the breaking of one Si-O bond is necessary for the motion of silicon atoms from one site to a neighbouring unoccupied site in the lattice.

The Si diffusion results challenge the widely accepted idea that oxygen diffusion controls climb, assumed to be the controlling mechanism for creep. Still the p_{O_2} dependence of creep indicates that the link has to exist between high temperature creep and diffusion, but now diffusion of silicon. Some indications have been given in the last section concerning our own opinion on this subject. The purpose of this paper is however merely to bring some hard facts into this debate.

References

Buening, D.K., and P.R. Buseck, FeMg lattice diffusion in olivine, J.G.R., 78, 6852, 1973.

Durham, W.B., Plastic flow of single crystal olivine, Ph. D. thesis, M.I.T., Boston, Mass., 1975.

Durham, W.B., and C. Goetze, Plastic flow of oriented single crystals of olivine, I mechanical data, J.G.R., 82, 5737, 1977.

Durham, W.B., C. Froidevaux, and O. Jaoul, Transient and steady-state creep of pure forsterite at low stress, Phys. Earth Planet. Inter., 19, 263, 1979.

Froidevaux,C., and G. Schubert, Plate motion and structure of the continental asthenosphere: a realistic model of the upper mantle, J.G.R., 80, 2553, 1975.

Gueguen, Y., High temperature olivine creep: evidence for control by edge dislocations, Geophys. Res. Letters, 6, 357, 1979.

Jaoul, O., M. Michaut, Y. Gueguen, and D. Ricoult, Decorated dislocations in forsterite, Phys. Chem. Minerals, 5, 15, 1979.

Jaoul, O., C. Froidevaux, W.B. Durham, and M. Michaut, Oxygen self-diffusion in forsterite: implications for the high temperature creep mechanism, Earth and Planet. Sc. Lett., 47, 391-397, 1980.

Kohlstedt, D.L., C. Goetze, W.B. Durham, and J.B. Vander Sande, A new technique for decorating dislocations in olivine, Science, 191, 1045, 1976.

Lager, G.A., and E.P. Meagher, High temperature structural study of six olivines, American Mineralogist, 63, 365, 1978.

Michaut, M., Fluage de la forstérite à haute température, thèse, Orsay, France, 1980.

Pluschkell, W., and H.J. Engell, Ionen und Elektronen Leitung im Magnesiummothosilikat, Ber. Dtsch. Keram. Ges., 45, 388, 1968.

Poirier, J.P., and B. Vergobbi, Splitting of dislocations in olivine, cross-slip controlled creep and mantle rheology, Phys. Earth Planet. Inter., 16, 370, 1978.

Reddy, K.P.R., and A.R. Cooper, Oxygen self-diffusion in forsterite and $MgAl_2O_4$, Am. Ceram. Soc. Bull., 57, 310, 1978.

Ross, J.V., H.G. Ave'Lallemant, and N.L. Carter, The activation volume for creep of olivine, Trans. Am. Geophys. Union, 59, 374, 1978.

Slodzian, G., Some problems encountered in secondary ion emission applied to elementary analysis, Surface Science, 48, 161, 1975.

Stocker, R.L., and M.F. Ashby, On the rheology of the upper mantle, Rev. Geophys. Space Phys., 11, 391, 1973.

Stocker, R.L., Influence of oxygen pressure on defect concentrations in olivine with a fixed cationic ratio, Phys. Earth Planet. Inter., 17, 118, 1978.

Weertman, J., The creep strength of the earth's mantle, Rev. Geophys. Space Phys., 8, 145-168, 1970.

Weertman, J., and J.R. Weertman, High temperature creep of rock and mantle viscosity, in Annual Reviews Inc., Palo Alto, California, 293-315, 1975.

EFFECT OF OXYGEN PARTIAL PRESSURE ON THE CREEP OF OLIVINE

D.L. Kohlstedt and Paul Hornack

Department of Materials Science and Engineering

Cornell University, Ithaca, New York 14853

Abstract. Single crystals of San Carlos peridot, $(Mg_{0.9}Fe_{0.1})_2SiO_4$, were deformed in one atmosphere of CO_2 plus H_2 with mixing ratios in the range 10:90 to 80:20 to study the effect of oxygen partial pressure on the creep behavior of olivine. In all of the experiments, the differential stress was held constant at 100 MPa, and the crystals were oriented so that the load axis was the $[101]_c$. Three sets of creep experiments were performed. (1) Constant $CO_2:H_2$ ratio: The creep rate was measured as a function of temperature, $1250 < T < 1400°C$, for two CO_2-H_2 mixtures, 70:30 and 20:80. For a given temperature, the creep rate in the more H_2-rich gas is $10^{0.5}$ smaller than in the CO_2-rich mixture. The activation energy for creep in either gas mixture is ∿536 kJ/mole. (2) Constant P_{O_2} : For a fixed mixing ratio, the oxygen partial pressure is a function of temperature. Therefore, temperature-cycling experiments were performed at fixed oxygen partial pressures by adjusting the CO_2-H_2 ratio at each temperature. At a constant oxygen partial pressure of 10^{-6}Pa, the creep activation energy is ∿448 kJ/mole. (3) Constant T: At 1322°C, several runs were made in which the oxygen partial pressure was varied over the range 10^{-4} to 10^{-7} Pa. The strain rate is proportional to ${P_{O_2}}^{1/6}$. Thus, the flow law for iron-bearing olivine under the conditions explored is of the form $\dot{\varepsilon} \propto {P_{O_2}}^{1/6} \exp(-\frac{448 \text{ kJ/mole}}{RT})$.

Introduction

For many crystalline solids, the rate of plastic deformation, $\dot{\varepsilon}$, at high temperatures (usually, $T > T_m/2$) appears to be limited by the rate of diffusion of the slowest moving species, S, such that

$$\dot{\varepsilon} \propto D_S \qquad (1)$$

where D_S is the diffusivity of the slowest diffusing species (see for example, Kirby and Raleigh, 1973; Takeuchi and Argon, 1976). This observation is the basis for mechanistic models of high-temperature creep in which the critical step is the climb of edge dislocations to annihilate other edge dislocations, to form low-angle boundaries or to circumvent obstacles (see for example, Weertman, 1972). The diffusivity of a given species is, in turn, proportional to the concentration of vacancies if diffusion occurs by a vacancy mechanism;

$$D_S \propto [V_S] \qquad (2)$$

where $[V_S]$ is the concentration of vacancies of the slowest diffusing species. Thus from Eqs. 1 and 2, the strain rate should be directly proportional to the concentration of vacancies (Shewmon, 1963),

$$\dot{\varepsilon} \propto [V_S]. \qquad (3)$$

(If diffusion occurs by an interstitial mechanism, then $\dot{\varepsilon} \propto [S_i]$, where $[S_i]$ is the concentration of the slowest diffusing interstitial ions.) If the critical microscopic step is cross slip of screw dislocations or glide of edge dislocations rather than diffusion-controlled climb of edge dislocations, the strain rate is still likely to depend upon the concentration of point defects which can act as obstacles. Charged point defects can be particularly effective obstacles because their electrostatic fields, which interact with dangling charged bonds or charged jogs along dislocation lines, are long ranged. Climb of edge dislocations (Kohlstedt and Goetze, 1974; Durham et al., 1977), glide of edge dislocations (Durham et al., 1977; Gueguen, 1979), and cross slip of screw dislocations (Poirier and Vergobbi, 1978; Zeuch and Green, 1979) have all been proposed as the rate determining step for high-temperature creep of olivine.

The equilibrium concentrations of point defects depend on the thermodynamic state of the solid. From the Gibbs phase rule, four independent thermodynamic state variables are required to determine the concentrations of point

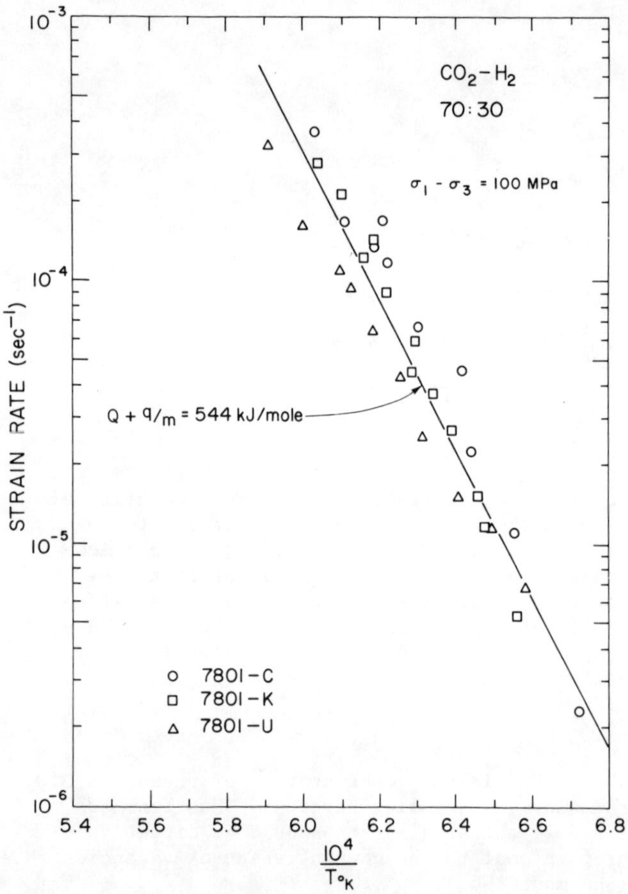

Fig. 1. Strain rate versus inverse temperature for three crystals deformed at a differential stress of 100 MPa in one atmosphere of CO_2:H_2 = 70:30.

defects in ternary, non-metallic crystals. For oxides, such as forsterite, pressure P, temperature T, oxygen partial pressure P_{O_2}, and activity of a coexisting neighboring phase (such as enstatite or periclase) are commonly chosen as the four independent state variables (Schmalzried, 1974, p. 44). In quaternary systems such as iron-bearing olivine, five state variables must be specified to define the thermodynamic state; and if water is present, an additional state variable (such as the activity of water) must be specified. Published studies of creep of olivine have, in general, been undertaken with control of only two (P and T) of the five or six independent state variables.

In the present paper, the effect of changes in the oxygen partial pressure on the high-temperature creep behavior of iron-bearing olivine single crystals at one atmosphere total pressure is explored. Variations in the concentrations of point defects with changes in oxygen partial pressure have been derived by Stocker (1978) for this partially closed system with fixed cation/cation ratios; in general,

$$[V_S] \propto P_{O_2}^{\ 1/m}. \tag{4}$$

Thus from Eqs. 3 and 4,

$$\dot{\varepsilon} \propto P_{O_2}^{\ 1/m}. \tag{5}$$

The long range goal of experimental studies such as the one described here is to provide insight into the microscopic mechanisms important in grain-matrix deformation. The results reported are a summary of the data of Hornack (1978) which were described briefly by Hornack and Kohlstedt (1979).

Experimental Details

Creep Apparatus

All experiments were performed inside a vacuum-tight, water-cooled stainless-steel chamber 15 cm in diameter and 15 cm high. The

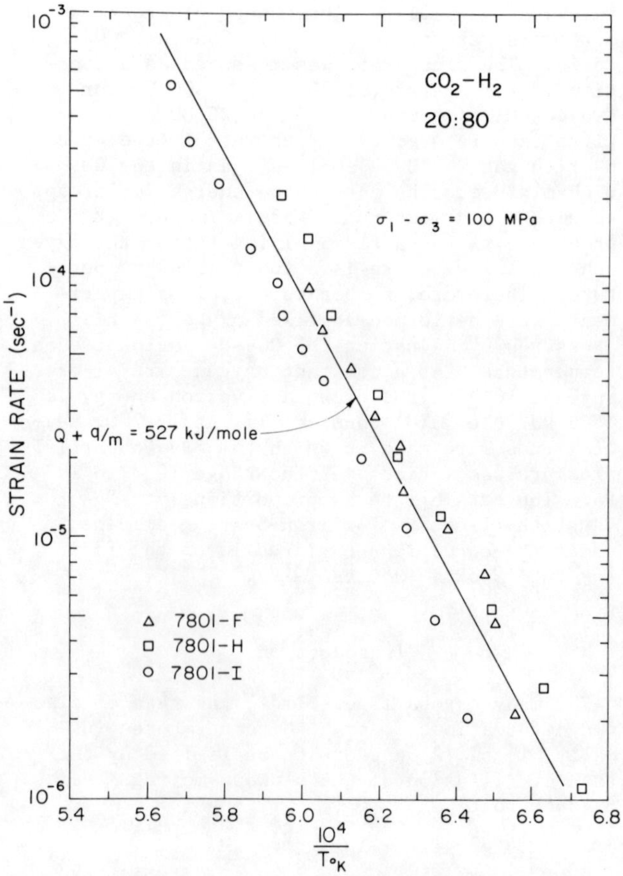

Fig. 2. Strain rate versus inverse temperature for three crystals deformed at a differential stress of 100 MPa in one atmosphere of CO_2:H_2 = 20:80.

deformation apparatus was similar to that described by Kohlstedt and Goetze (1974) and Durham and Goetze (1977). After a specimen was positioned for a deformation test, the chamber was evacuated to $\sim 10^{-2}$ Torr and then back-filled to one atmosphere with the appropriate CO_2-H_2 gas mixture. The gas, which flowed through an axial hole in the bottom piston, was released into the chamber about 0.5 cm from the specimen. With this design, the gas was heated before reaching the sample and the atmosphere near the specimen had the appropriate CO_2-H_2 ratio. A flow rate of ~ 10 ml sec^{-1} was maintained. The gas escaped through a leaky seal along the top piston.

The deformation column consisted of two 1 cm diameter, tungsten - 2% thoria pistons. One piston was rigidly attached to the bottom of the test chamber; the other, guided by two ball races each 2 cm long, entered through the top cover of the chamber. A guide pin prevented rotation of the top piston. The olivine samples did not indent the thoriated-tungsten pistons at a differential stress of 100 MPa for the test temperatures of 1200 to 1500°C used in this study. Thus, the dispersion-strengthed material is significantly stronger than the pure tungsten used previously (Kohlstedt et al. 1976; Durham and Goetze, 1977). The thoriated tungsten also appears to oxidize less readily than pure tungsten, so that the friction between the polished ends of the pistons and the polished surfaces of the sample remained low during the entire experiment.

The samples were deformed in compression by dead-weight loading the upper piston. Two direct current differential transformers (DCDT's), mounted diametrically opposite each other, measured the position of the weight pan relative to the test chamber and, thus, provided a continuous measure of the length of the sample. The summed DCDT output voltage was recorded on a strip-chart recorder. Strain rates calculated from these data were accurate to ±5%.

Furnaces

To explore a wide range of oxygen partial pressure, three types of furnace element were employed. For the most oxidizing gas mixtures (greater than 70% CO_2), Pt-40% Rh wire was wound on an Al_2O_3 tube. The heating element for gas mixtures with 30 to 70% CO_2 was tungsten wire, again wound on an Al_2O_3 tube. For more reducing mixtures, a thin-walled tube of molybdenum was supported between graphite rings. Two thermocouples contacted the specimen in each furnace design; one provided input for the temperature controller, the other input for a chart recorder. Thermocouples of Pt/Pt-13% Rh were used in the most oxidizing conditions; thermocouples of W-5% Re/W-26% Re were used otherwise. The temperature varied by less than 5°C from one end of a specimen to the other. In experiments in which the temperature was cycled, relative temperatures were accurate to ±5°C.

TABLE 1. Activation energies determined for individual samples and total strain at end of run.

Sample Number	Activation Energy	Comments	Total Strain
7801-C	544 kJ/mole	CO_2:H_2 = 70:30	21.6%
7801-K	607 "	" "	25.1%
7801-U	481 "	" "	24.2%
7801-F	485 "	CO_2:H_2 = 20:80	15.7%
7801-H	527 "	" "	12 %
7801-I	569 "	" "	25 %
7801-Z	494 "	$P_{O_2} = 10^{-6}$ Pa	13.3%
7801-a	460 "	" "	8.0%
7801-d	439 "	" "	8.5%
7801-f	452 "	" "	9.2%

After tests in gas mixtures with more than 50% CO_2, the thoriated-tungsten pistons were covered with a thin oxide layer. The displacement-time records were slightly jerky, apparently due to specimen-piston friction.

Samples

All specimens were cut from one single crystal of San Carlos peridot. The specimens were oriented by the Laue back-reflection method so that the load axis was within 2° of the cubic [101] direction, i.e., 45° from [100] and 45° from [001]. The specimens were square prisms with approximate dimensions of 4x2x2 mm^3. The small ends were ground flat and parallel to within 0.001 cm and then polished to a 1 μm finish.

Test Conditions

All data reported here were obtained at differential stresses between 95 and 100 MPa. Weights were added during the creep tests, under the assumption that the sample deformed homogeneously, to maintain this stress level. All strain rates were corrected to a differential stress of 100 MPa using the relation $\dot{\varepsilon} \propto \sigma^{3.6}$, the empirical relation reported for creep of single crystals with the $[101]_c$ orientation (Durham and Goetze, 1977).

Three types of creep tests were performed: (1) One group of samples were deformed in one atmosphere of CO_2-H_2 at a fixed mixing ratio of either 70:30 or 20:80 over the temperature range 1200 to 1500°C. (2) A second group of samples were deformed at a fixed oxygen partial pressure of 10^{-6} Pa over the temperature range 1230 to 1450°C. (3) A third set of samples were deformed at 1322°C over the P_{O_2} range $10^{-7.5}$ - 10^{-5} Pa.

Experiments at fixed CO_2:H_2 ratios

This first group of experiments was of an exploratory nature to determine whether a change in oxygen partial pressure affected the creep rate of olivine. Strain rate as a function of inverse temperature is plotted in Fig. 1 for three crystals deformed in 70:30 mixtures of CO_2-H_2 at a

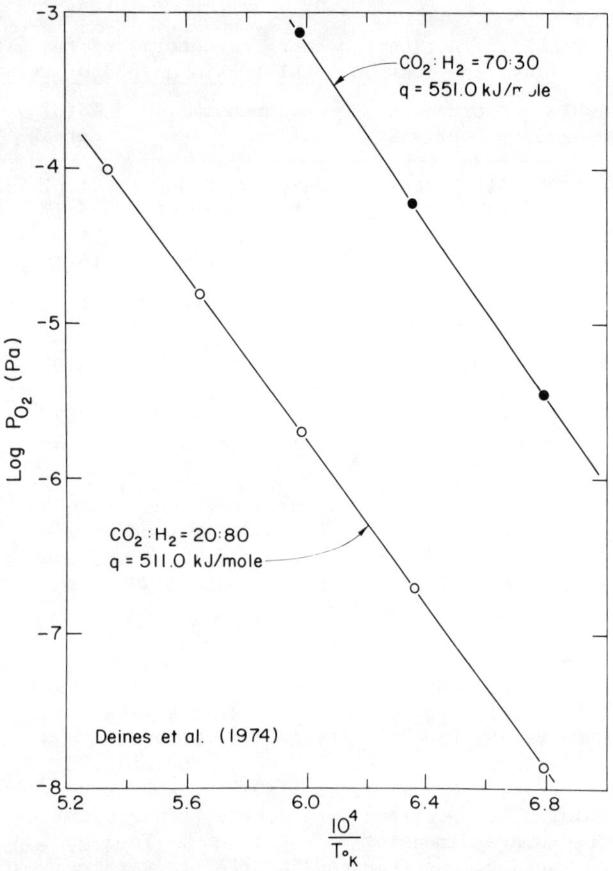

Fig. 3. Log P_{O_2} versus inverse temperature based on tabulation of Deines et al. (1974).

differential stress of 100 MPa; the data were obtained during at least two heating and cooling cycles. Similar data are plotted in Fig. 2 for three crystals crept in 20:80 mixtures of CO_2-H_2. Each datum point represents at least 1% strain. A new steady state creep rate was reached within 15 minutes after changing the temperature. The average activation energies for the three samples deformed in the 70:30 CO_2-H_2 gas mixture and for the three samples deformed in the 20:80 mixture are 544 ± 63 kJ/mole and 527 ± 43 kJ/mole, respectively. The activation energies and total strain for the individual runs are listed in Table I. For these constant stress experiments (σ_1-σ_3 = 100 MPa), at a given temperature the rate of deformation of the olivine crystals increased as the CO_2:H_2 ratio increased.

For a given CO_2:H_2 ratio, the oxygen partial pressure is a function of temperature (Muan and Osburn, 1965; Deines et al., 1974). Log P_{O_2} is plotted versus inverse temperature for the 70:30 and 20:80 CO_2-H_2 mixtures in Fig. 3. Least squares fitting of these data to straight lines yield

$$P_{O_2}\big|_{70:30} = 6.9\text{x}10^{-1}\exp\left(-\frac{551\ \text{kJ/mole}}{RT}\right)\text{Pa} \quad (6)$$

and

$$P_{O_2}\big|_{20:80} = 2.3\text{x}10^{4}\exp\left(-\frac{511\ \text{kJ/mole}}{RT}\right)\text{Pa}. \quad (7)$$

In general,

$$P_{O_2}\big|_{CO_2:H_2} = K\ \exp\ (-q/RT) \quad (8)$$

where K and q depend upon the CO_2:H_2 ratio. Thus if the strain rate depends upon oxygen partial pressure, the activation energy measured in creep experiments performed in one atmosphere of fixed CO_2:H_2 ratio will contain a contribution due to the temperature dependence of the oxygen partial pressure. Restated, if from Eqs. 1 and 5

$$\dot{\varepsilon} \propto P_{O_2}^{\ 1/m}\ \exp\ (-Q/RT), \quad (9)$$

then from Eq. 8

$$\dot{\varepsilon}\big|_{CO_2:H_2} \propto \exp\ -\left[\left(\frac{q/m + Q}{RT}\right)\right] \quad (10)$$

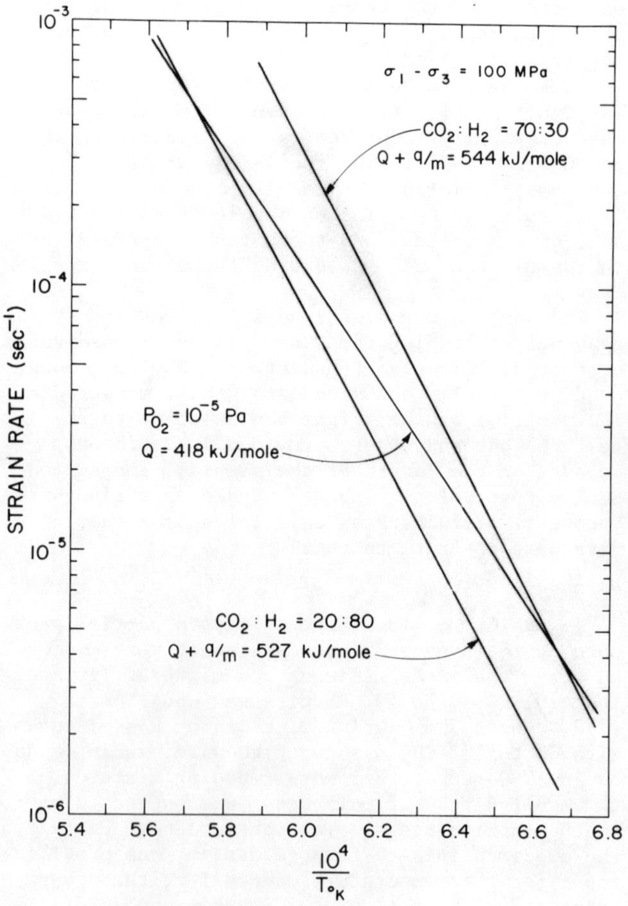

Fig. 4. Summary of strain rate versus inverse temperature results CO_2:H_2 = 70:30 (Fig. 1) and CO_2:H_2 = 20:80 (Fig. 2). A line connecting the point at $P_{O_2} = 10^{-5}$ Pa on the 70:30 curve to the point at $P_{O_2} = 10^{-5}$ Pa on the 20:80 curve is also shown.

where m depends upon details of the point-defect chemistry and Q is the activation energy for creep at fixed P_{O_2}.

An estimate of strain rate as a function of temperature at constant P_{O_2} can be made by selecting the strain rate - temperature point at a given P_{O_2} level from each set of $\dot{\varepsilon}$ - 1/T data at constant mixing ratio (i.e., from Figs. 1 and 2). To illustrate this point, the strain rate - temperature results from Figs. 1 and 2 are summarized in Fig. 4 with the addition of the strain rate - temperature curve at a P_{O_2} of 10^{-5} Pa. The slope of the constant P_{O_2} curve yields an activation energy of 418 kJ/mole. A value for m in Eq. 9 can now be calculated by combining the activation energy for creep at constant mixing ratio (Eq. 10, q/m + Q), with that at constant P_{O_2} (Eq. 9, Q) and the temperature dependence of P_{O_2} at fixed $CO_2:H_2$ ratio (Eq. 8, q); $3 \leq m \leq 12$.

Experiments at fixed P_{O_2}

In a second series of experiments, four single crystals were deformed at an oxygen partial pressure of 10^{-6} Pa over the temperature range 1230 to 1450°C at a differential stress of 100 MPa. The data are presented in Fig. 5 on a plot of strain rate versus inverse temperature. Data were collected during at least two heating/cooling runs on each sample. Each datum point represents at least 1% strain. No transient creep was observed after a change in temperature. (Because the DCDT's are external to the furnace, it takes approximately five minutes before the chart record is free of effects due to thermal expansion or contraction of the loading column.) The average activation energy for the four runs is 448 ± 54 kJ/mole. The activation energies and total strain for the individual runs are listed in Table I.

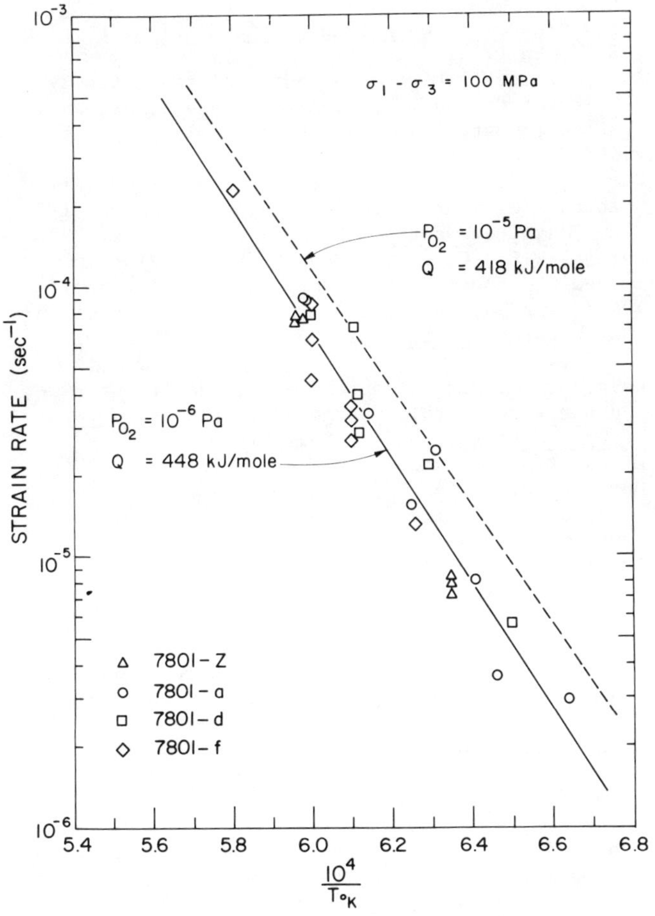

Fig. 5. Strain rate versus inverse temperature for four samples deformed at P_{O_2} = 10^{-6} Pa and a differential stress of 100 MPa. The dashed line at P_{O_2} = 10^{-5} Pa is taken from Fig. 4.

The strain rate - inverse temperature curve at P_{O_2} = 10^{-5} Pa from Fig. 4 is also shown in Fig. 5. At a given temperature, the separation between these lines along the strain-rate axis yields $3 \leq m \leq 6$ (see Eq. 9). A value for m can also be calculated from the measured activation energy combined with the activation energy for creep at fixed $CO_2:H_2$ ratio (using Eqs. 8,9, and 10); $4 \leq m \leq 20$.

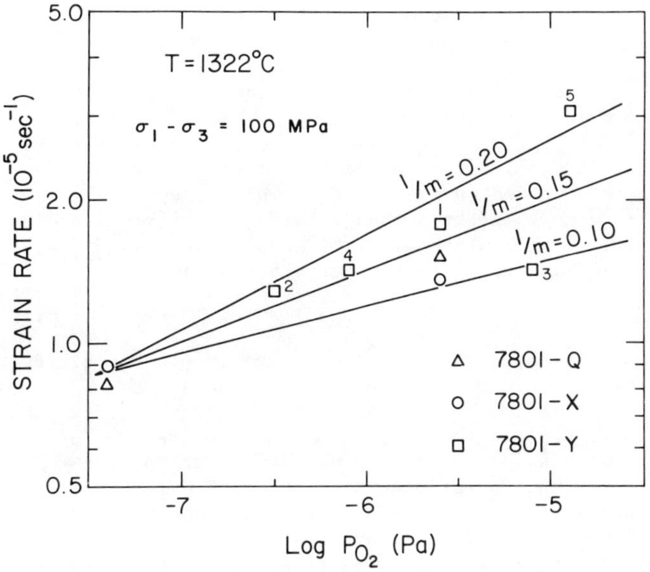

Fig. 6. Strain rate versus log P_{O_2} for three samples deformed at 1322°C and a differential stress of 100 MPa. The order of changes in P_{O_2} is noted for sample 7801-Y.

Experiments at fixed temperature

In a third series of experiments, three crystals were deformed at 1322°C over the range of oxygen partial pressures $10^{-7.4} \leq P_{O_2} \leq 10^{-4.9}$ Pa at a differential stress of 100 MPa. The data from these runs are plotted in Fig. 6 as strain rate versus log P_{O_2}. The slopes of straight lines fitted to the data for the individual samples yield $4 \leq m \leq 10$. From Eq. 10, this range of values for m combined with the activation energy measured in an atmosphere of fixed $CO_2:H_2$ ratio yields an activation energy for creep Q = 439 ± 42 kJ/mole.

In runs 7801-Q and 7801-X, the sample was deformed 3% at the lower P_{O_2} and then 3% at the higher P_{O_2}. In run 7801-Y, the sample was deformed at least 1% at each P_{O_2}; the datum points are numbered in the order in which they were obtained. After a change in P_{O_2} (at fixed T and $\sigma_1-\sigma_3$), a new steady state strain rate was reached after approximately 15 minutes.

Discussion

The three types of creep experiments described above yield an average value of $\langle m \rangle = 6$ and $\langle Q \rangle = 435$ kJ/mole. This activation energy is 88 kJ/mole smaller than that reported for creep of olivine single crystals in one atmosphere of constant $CO_2:H_2$ ratio (Kohlstedt and Goetze, 1974; Kohlstedt et al., 1976; Durham and Goetze, 1977). Jaoul et al. (1981) have also recently reported m=6 for creep of iron-bearing olivine crystals at one atmosphere confining pressure. Thus, the flow law for olivine single crystals should be of the form

$$\dot{\varepsilon} = f(\sigma)\, P_{O_2}^{1/6} \exp(-\frac{448 \text{ kJ/mole}}{RT}) \sec^{-1}. \qquad (11)$$

where $f(\sigma)$ is an empirical function of differential stress (see, for example, Kohlstedt et al., 1976, Fig. 1). A $P_{O_2}^{1/6}$ dependence has also been reported for the interdiffusion coefficients of Fe and Mg in olivine under the condition of a partially closed system (Buening and Buseck, 1973).

Systematic changes of P_{O_2} have an important influence on the flow law for iron-bearing olivine. No attempt has been made here to explore the effects of varying the Mg/Fe ratio or the (Mg + Fe)/Si ratio. In comparing the creep data on olivine from several laboratories, attention has not been paid to the thermodynamic states under which the samples were deformed and differences have been loosely assigned to "water weakening" effects. It is suggested here that differences in strengths observed from one laboratory to the next may, at least in part, be due to differences in concentrations in point defects. For example, in the experiments at one atmosphere on single crystals, the thermodynamic state -- and thus the concentrations of point defects -- is fixed by the P_{O_2} of the buffering gas plus the as-received Mg/Fe and (Mg + Fe)/Si ratios (as well as P and T). In contrast, for experiments in a solid-medium apparatus, the confining medium (e.g., AlSiMag or talc) will set the thermodynamic variables (other than P and T). Lack of control of thermodynamic variables resulted in enormous scatter in the electrical conductivity versus temperature data for olivine (Duba, 1976); possibly some of the scatter in the creep data has a similar origin.

No discussion has been presented above of the influence of extrinsic point defects on the mechanical properties of olivine. The introduction of extrinsic defects, such as OH^-, will have a major effect on the concentrations of other point defects, such as oxygen vacancies (Hobbs, 1979). Thus, the extrinsic defects can drastically alter the flow rate of olivine by increasing, for example, the diffusivity of the slowest diffusing species (see Eqs. 1 and 2) or by changing the obstacle structure which limits dislocation motion.

Acknowledgements. This research was supported by the National Science Foundation through Grant No. EAR7919725. The high-temperature furnace used in this research was part of the Materials Science Center's central facility for crystal growth.

References

Buening, D.K. and P.R. Buseck, Fe-Mg lattice diffusion in olivine, J. Geophys. Res., 78, 6852-6862, 1973.

Deines, P., R.H. Nafziger, G.C. Ulmer, and E. Woermann, Temperature-oxygen fugacity tables for selected gas mixtures in the system C-H-O at one atmosphere total pressure, Bull. Earth and Mineral Sciences Experiment Station, Penn. State Univ., 1974.

Duba, A., Are laboratory electrical conductivity relevant to the earth?, Acta Geodaet., Geophys. et Montanist. Acad. Sci. Hung. Tomus, 11, 485-495, 1976.

Durham, W.B. and C. Goetze, Plastic flow of oriented single crystals of olivine: 1. Mechanical data, J. Geophys. Res., 82, 5737-5753, 1977.

Durham, W.B., C. Goetze, and B. Blake, Plastic flow in oriented single crystals of olivine: 2. Observations and interpretations of dislocation structures, J. Geophys. Res., 82, 5755-5770, 1977.

Gueguen, Y., High-temperature olivine creep: Evidence for control by edge dislocations, J. Geophys. Letters, 6, 357-360, 1979.

Hobbs, B.E., private communication, 1979.

Hornack, P.G., The Effect of Oxygen Fugacity on the Creep of Olivine, M.S. Thesis, Cornell University, 99 pp., 1978.

Hornack, P. and D.L. Kohlstedt, The Effect of oxygen partial pressure on the creep of olivine, EOS, 60, 369-370, 1979.

Jaoul, O., C. Froidevaux, W.B. Durham, and M. Michaut, Oxygen self-diffusion in forsterite: Implications for high-temperature creep mechanism, preprint (1981).

Kirby, S.H. and C.B. Raleigh, Mechanisms of high-temperature, solid state flow in minerals and ceramics and their bearing on creep behavior of the mantle, Tectonophysics, 19, 165-194, 1973.

Kohlstedt, D.L. and C. Goetze, Low-stress high-temperature creep in olivine single crystals, J. Geophys. Res., 79, 2045-2051, 1974.

Kohlstedt, D.L., C. Goetze, and W.B. Durham, Experimental deformation of single crystal olivine with application to flow in the mantle, in Petrophysics: The Physics and Chemistry of Minerals and Rocks, edited by R.G.J. Strens, John Wiley and Sons, Ltd., London, pp. 35-49, 1976a.

Muan, A. and E.F. Osburn, Phase Equilibria Among Oxides in Steel Making, Addison-Wesley, Reading, Mass., p. 50, 1965.

Poirier, J.P. and B. Vergobbi, Splitting of dislocation in olivine, cross-slip controlled creep and mantle rheology, Phys. Earth Planet. Inter., 16, 370-379, 1978.

Schmalzried, H., Solid State Reactions, Academic Press, New York, 212 pp., 1974.

Shewmon, P.G., Diffusion in Solids, McGraw-Hill, New York, 1963.

Stocker, R.L., Influence of oxygen pressure on defect concentrations in olivine with a fixed cationic ratio, Phys. Earth and Planet. Inter., 17, 118-129, 1978.

Takeuchi, S., and A.S. Argon, Steady-state creep of single-phase crystalline matter at high temperature, J. Mater. Sci., 11, 1542-1566, 1976.

Weertman, J., High-temperature creep produced by dislocation motion, in J.E. Dorn Memorial Symposium, Cleveland, 1972.

Zeuch, D.H. and H.W. Green, Experimental deformation in an "anhydrous" synthetic dunite, Bull. Mineral., 102, 185-187, 1979.

The Influence of Strain Rate and Moisture Content on Rock Failure

H. Spetzler

Department of Geological Sciences and
Cooperative Institute for Research in
Environmental Sciences
University of Colorado/NOAA
Boulder, CO 80309

C. Sondergeld and I. C. Getting

Cooperative Institute for Research in
Environmental Sciences
University of Colorado/NOAA
Boulder, CO 80309

Abstract

Failure in brittle rock occurs by the interaction of small cracks. The strength and mechanical behavior of a rock of given mineralogy and grain size depends upon the size, shape, and distribution of small cracks. Cracks develop in response to applied stress in a manner dependent upon the moisture content at the crack tips. If stress is applied at a low rate and moisture is abundant at the crack tips, cracks will grow to be large and coalesce into a failure plane at low stress; i.e. the rock is weak. If stress is applied rapidly and the moisture content is low, the limited growth of the existing cracks will not relieve the stress concentrations and many new small cracks will be formed. In this case failure occurs by the coalescence of a larger number of smaller cracks and the rock is strong. We assume that for a given crack distribution and morphology (i.e. size and shape of the cracks) the rock will fail when the cracks have reached an average critical length and that the crack growth is governed by an activated mechanism (in this case stress corrosion by water). These assumptions permit one to link crack growth data to failure strength (S_f) through the following equation:

$$S_f = S_u\left[1-\frac{TR}{Q^*}[\ln T + n\ln P - \ln\dot{S} + B\,]\right]$$

where $\dot{S}$ is the applied stress rate, T the temperature, P the partial pressure of water, n the order of the chemical reaction ($n=1$ for our experiments), R the universal gas constant and Q^* the activation energy for stress corrosion. S_u and B, which must be determined experimentally, are constants which depend upon rock type, activation volume, and initial crack distribution and morphology.

Experimental results from single crack propagation in thin plates of glass and rock yield activation energy, activation volume, crack velocity and stress intensity factor. Triaxial deformation experiments give strength and suggest that the crack morphology depends upon stress rate and moisture content.

Introduction

The application of increasing nonhydrostatic stress to rocks results in the growth of previously existing cracks and in the formation and growth of new cracks. Failure occurs when the cracks coalesce into a major fracture and the rock can no longer sustain the applied deviatoric stress (for example see, Scholz, 1972; Martin, 1972) In the presence of moisture, crack growth may be dominated by stress corrosion at the crack tip.

A number of authors (Tapponnier and Brace, 1976; Siegfried and Simnons, 1978; Kranz, 1979; Swanson, 1979) have found distinct crack types in rocks which were clearly dependent upon the loading history of the sample. Swanson (1979) showed that the ratio of transgranular (through grains) to intergranular (around grain boundaries) crack lengths were different by a factor of 2 (from 0.6 to 1.2) when thin plates of Westerly granite were deformed under the extremes of his experimental conditions. Low humidity environments and fast loading rates enhance transgranular cracking. Swanson interprets his results in terms of the stress corrosion theory developed for glasses by Charles and Hillig (1965) and as extended by Wiederhorn (1969). (For a review and applications to geological problems see Anderson and Grew, 1977.) Figure 1 shows the three regions of stable crack growth: Region I where the rate of crack growth is governed by stress corrosion at the crack tip in equilibrium with

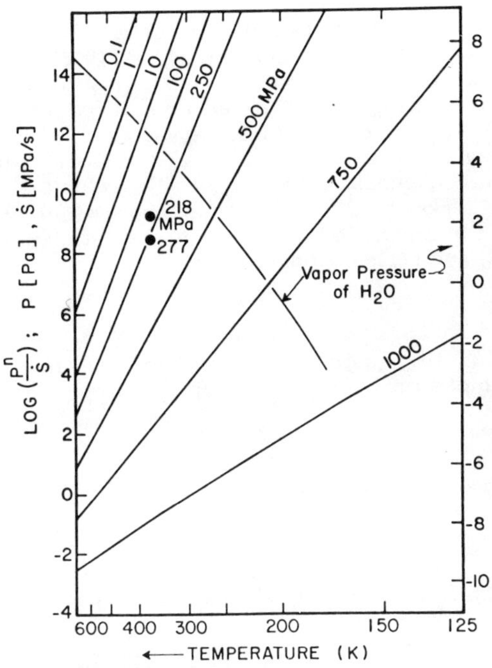

Fig. 1. (from Swanson (1979)). Log crack velocity versus stress intensity factor, K. Above K_C, crack propagation is unstable.

the environment; Region II where the rate of crack growth is limited by the transport of water vapor, i.e., the environment of the crack tip is not in equilibrium with the surrounding environment and Region III where crack growth is independent of the presence of water vapor. The differences in the crack morphology observed by Swanson can be attributed in part to the mechanisms of crack growth which prevail in Region I for the slow moist experiment and to the mechanisms of Region III for the fast dry experiment.

Scholz and Koczynski (1979) attribute strain rate dependent deformation, observed during stress cycling experiments in Westerly granite, to specific crack morphologies developed during the loading history of the rock. Brodsky and Spetzler (1979) find strain rate and moisture dependent "elastic constants" in Ralston basalt well below the onset of dilatancy. They, too, appeal to the dependence of crack morphologies upon moisture and strain rate to explain their results. Here again, at least part of the differences in the crack populations may be attributable to crack growth spanning from Region I to Region III.

Failure Criterion

In the following we wish to examine the effect of crack growth in Region I on the strength of rock. We assume: (a) that at the beginning of the experiment the crack morphology (i.e. crack shape, length) and distribution within a specimen are only dependent upon of the rock type and its previous loading history, (b) that failure of the rock occurs by the coalescence of cracks that have grown in Region I and (c) that the coalescence occurs when those cracks that participate in failure have grown by a critical amount. This differs from our earlier approach (Mizutani et al., 1977; and Soga et al., 1979) where we considered the strain rate at failure to be directly related to the crack velocity.

The crack growth equations used in the ceramics literature in describing single cracks (e.g. Wiederhorn, 1967) are of the form

$$\dot{a} = cP^n e^{-\frac{Q^*-V^*\sigma}{RT}} \tag{1}$$

where

$\dot{a}$ = velocity of crack growth

P = partial pressure of water at the crack tip

Q^* = activation energy for the stress corrosion reaction

V^* = activation volume for the stress corrosion reaction

σ = tensile stress at the crack tip

T = temperature

R = gas constant = 8.3 J/mole · K

c = constant

n = constant, determining the order of the chemical reaction

For geological applications we must consider the limit as the stress approaches zero. In this case the value of the crack velocity in equation (1) becomes a constant. Figure 2 shows the energy diagram illustrating both the forward and backward reactions. Without stress at the crack tip the number of bonds that are broken is equal to the number healed and the crack velocity is zero. To apply to the very low stress case, equation (1) must be modified to include a healing term:

$$\dot{a} = cP^n \left[e^{-\frac{Q^*-V^*\sigma}{RT}} - e^{-\frac{Q^*}{RT}} \right] \tag{2}$$

The first term in brackets represents bond breaking; the second, bond healing.

According to the above mentioned criterion, namely that the rock fails when the cracks participating in failure have grown by an amount K, failure occurs when

$$\int_0^{t_f} \dot{a} dt = K \tag{3}$$

where $\dot{a}$ now applies to typical crack velocities. The integration is over time, and starts at a time before which all samples that are to be compared have had a common history; i.e. at t=0 they have a common crack morphology and distribution. The integration continues until the time of failure. The crack growth is assumed to remain in Region I. While the failure criterion should work for creep and cycling experiments as well, here we will develop it for laboratory and field situations where the rate of stress increase is constant:

$$S = \dot{S}t \tag{4}$$

where S the applied deviatoric stress and $\dot{S}$ is the rate of stress increase. We assume that the average stress at the crack tips, σ, is linearly related to the applied stress such that

$$\sigma = \gamma S \tag{5}$$

γ being the stress magnification factor.

During a constant stress rate experiment the typical crack velocity at any time is

$$\dot{a} = cP^n e^{-\frac{Q^*}{RT}}\left[e^{\frac{\gamma V^* \dot{S}t}{RT}} - 1\right] \qquad (6)$$

Integrating (3) over time and substituting the failure strength S_f for $\dot{S}t_f$ yields:

$$\frac{cP^n RT}{\dot{S}\gamma V^*} e^{\frac{-Q^*}{RT}}\left[e^{\frac{\gamma V^* S_f}{RT}} - \frac{\gamma V^* S_f}{RT} - 1\right] = K \qquad (7)$$

Before applying equation (7) to real rocks let us evaluate the order of magnitude of the terms involved. It is clear from Figure 2 that when σV^* approaches Q^* the model must break down and failure must occur by a different mechanism. Therefore, we will consider only the case where

$$\sigma \leq \frac{Q^*}{V^*} \qquad (8)$$

Values for the activation energy Q^* and the term σV^* are determined experimentally by measuring crack velocity as a function of temperature and stress separately and then fitting the data to an equation similar to equation (1). Unfortunately the authors are not aware of a complete data set for any rocks. Q^* has been estimated by Soga et al., (1979) for Ralston basalt to be 110 kJ/mole (26 kcal/mole). Measurements on a glass with a composition similar to Ralston basalt yield an activation volume of 3.1×10^{-6} m^3/mole. Calculating the maximum stress at the crack tip according to (8) yields 35 GPa (350 kbars) which is within about a factor of two of the Youngs modulus of Ralston ~80 GPa. While this value is on the order of a theoretical bond strength and therefore somewhat encouraging, it depends strongly on the value of the activation volume which in turn depends upon estimates of the actual crack tip radii. Following Wiederhorn (1969), a crack tip radius of 5×10^{-10} meters was used. A more directly measured quantity in single crack propagation experiments is the product σV^* or its equivalent bK_I. K_I, the mode I stress intensity factor, is related to σ and the crack tip radius ρ by

$$\sigma = \frac{2K_I}{\sqrt{\pi\rho}} \qquad (9)$$

For Region I, the slope, b/RT, of a log crack velocity versus K_I graph has been determined for Westerly granite by Swanson and Spetzler (1979) and by Atkinson (1979). The former obtained a value of b = 0.13 $m^{5/2}$/mole; the latter, a value that is lower by nearly a factor two. The discrepancy is not yet understood but is being investigated. The maximum value of K_I, i.e. K_{IC} (see Figure 1), was measured in both laboratories and agrees very well, 1.8×10^6 N/$m^{3/2}$ and 1.74×10^6 N/$m^{3/2}$ respectively. Calculating the value of $\sigma V^* = bK_I$ we obtain 230 kJ/mole (55 kcal/mole) using the value of b from Swanson and Spetzler (1979) and 120 kJ/mole (28 kcal/mole) using b from Atkinson (1979). The activation energy Q^* for subcritical crack growth in Region I has not been measured for Westerly granite but is expected to be on the order of 100 kJ/mole. Values for Q^* for quartz have been reported in the literature as 46 to 100 kJ/mole by Scholz (1972), 63 ± 16 kJ/mole by Martin and Durham (1975) and 52.5 ± 3.8 kJ/mole by Atkinson (1979). Atkinson, as reported by Rudnicki (1979), also measured a value of 69.5 ± 1.7 kJ/mole for Arkansas Novaculite.

Having established that Q^* is of the same order of magnitude as maximum values of σV^*, we define a new value of stress σ_u at the crack tip, above which cracks cease to be subcritical and the representation in equations (2) and (7) is no longer valid.

$$\sigma_u = \frac{Q^*}{V^*} \qquad (10)$$

Analogous to equation (5) we define the maximum strength of a rock sample in compression as;

$$S_u = \frac{\sigma_u}{\gamma} \qquad (11)$$

The stress magnification factor, γ becomes:

$$\gamma = \frac{\sigma_u}{S_u} = \frac{Q^*}{S_u V^*} \qquad (12)$$

Using (1), (2), (3) and (5), the expression for the crack velocity (6) may be rewritten as:

$$\dot{a} = cP^n e^{-\frac{Q^*}{RT}}\left[e^{\frac{Q^* S}{RT S_u}} - 1\right] \qquad (13)$$

and expression (7) involving the failure strength as:

$$K = \frac{cP^n RTS_u}{\dot{S}Q^*} e^{\frac{-Q^*}{RT}}\left[e^{\frac{Q^* S_f}{RTS_u}} - \frac{Q^*}{RT}\frac{S_f}{S_u} - 1\right] \qquad (14)$$

We estimate the magnitude of c in equation (13) by using a value for Q^* of 110 kJ/mole and the data from single crack propagation experiments (Swanson, 1979). At 300 K, Swanson finds a crack tip velocity of $\sim10^{-5}$ m/s at a partial pressure of water of 2.7×10^{-3} Pa (20 torr) and $K_I/K_{Ic} = 2/3$. According to Martin (1972) and Soga et al (1979) the chemical reaction at the crack tip is a first order reaction; i.e., $n=1$. Using these values we find c = 1.24 m/Pa·s . To check if the

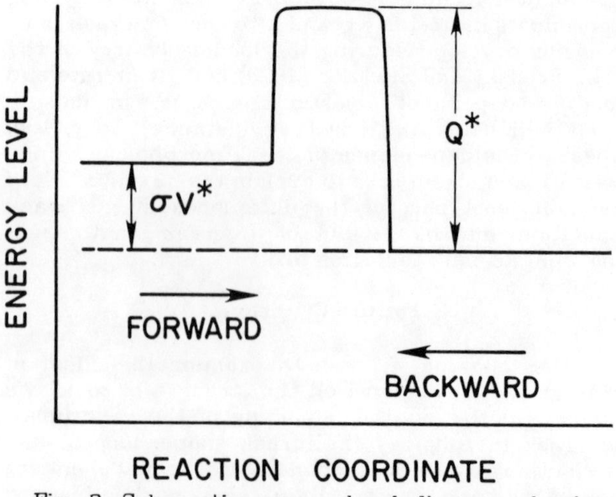

Fig. 2. Schematic energy level diagram showing the energy barriers for the forward and backward reaction for stress corrosion at crack tips.

TABLE 1

Sample No.	S_f (MPa)	$\dot{S}$ (MPa/s)	P (Pa) obs.	P (Pa) Calc.
10	392	2.4×10^{-5}	5.3×10^{-6}	2.3×10^{2}
11	218*	2.4×10^{-5}	4.0×10^{4}	
17	443	1.4×10^{-4}	2.7×10^{-6}	2.7×10^{2}
18	277*	1.4×10^{-4}	4.0×10^{4}	

Samples of Ralston basalt were deformed at T=363 K under the above conditions. The data in the right hand column give the effective moisture environment for samples 10 and 17 based upon equation (18).

*These data points were used to calculate S_u and C/K.

expected crack extension K is at all reasonable we integrate (13) and rewrite (14) to get:

$$K = \frac{\dot{a}RTS_u}{\dot{S}Q^*} \tag{15}$$

where $\dot{a}$ is now the crack velocity at the failure stress S_f. At a strain rate of 10^{-8}/s, unconfined Westerly granite has a strength of about 220 MPa (2.2 kbars). This makes $S_u = (3/2)\,S_f = 330$ MPa (3.3 kbars). Assuming failure occurs at a strain of 1%, the corresponding stress rate is 22 kPa/s (0.22 bars/s). Using these values the expected crack extension, K, is 3×10^{-3} m . This is a reasonable value and is on the same order as the grain size.

For $\frac{Q^*}{RT}\frac{S_f}{S_u} \gg 1$ equation (14) may be written as:

$$K = \frac{cP^nRTS_u}{\dot{S}Q^*}e^{-\frac{Q}{RT}(1-\frac{S_f}{S_u})} \tag{16}$$

or solving for $\frac{S_f}{S_u}$.

$$\frac{S_f}{S_u} = 1 - T\frac{R}{Q}\left[\ln T + n\ln P - \ln\dot{S} + \ln S_u + \ln\frac{R}{Q^*} + \ln\frac{C}{K}\right] \tag{17}$$

For an activation energy Q^* of 100 kJ/mole, the above inequality holds, for example, at 500 K for strength reductions S_f/S_u up to about ten; i.e. equations 16 and 17 are valid under those conditions. The constant $\ln(C/K)$ may be thought of as a measure of the crack morphology of the rock before the failure test.

Discussion

We will now examine some laboratory rock strength data and use equation (17) to extrapolate to geologic conditions. As Scholz and Koczynski (1979) point out, stress corrosion cracking is the dominant mode of cracking at low stress rates and for small stress amplitude cycling when moisture is present. Constant stress rate experiments on samples of Ralston basalt were performed by Mizutani et al. (1977). Their sample numbers 10, 11, 17 and 18 had a common history. They were baked for a period of approximately 100 hours at 670 K in a vacuum chamber at a pressure of 1.3×10^{-4} Pa (10^{-6}torr). All were broken at 363 K at various strain rates and moisture contents. Table 1 gives the conditions under which they were loaded. The last column gives the vapor pressure that was calculated for those samples deformed under high vacuum. The calculations are based upon the equation

$$S_f = 1230\left[1 - 7.55\times10^{-5}T\,(\ln T + n\ln P - \ln\dot{S} + 7.80)\right] \tag{18}$$

The data from samples 11 and 18 were used to calculate the ultimate strength S_u = 1230 MPa and the combination of constants

$$\left[\ln S_u + \ln\frac{R}{Q} + \ln\frac{C}{K}\right] = 7.80$$

The close agreement between the calculated values of partial pressure of water for samples 10 and 17 (see Table 1) gives us some confidence that the treatment of the samples was quite similar and that our approach toward the development of the strength equations (14) and (17) is basically correct. However, our confidence in being able to dry out samples is badly shaken. The pressure in the vacuum chamber was 8 orders of mag-

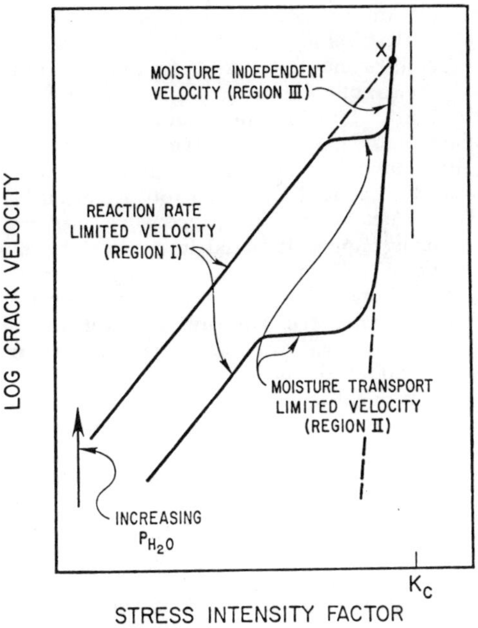

Fig. 3. Contour map of failure stress appropriate for Ralston basalt. The left ordinate is the logarithm of the ratio of vapor pressure of H_2O, P^n, to the rate at which stress is applied to the rock, $\dot{S}$. The right ordinate applies to a constant stress rate of 10^{-7} MPa/s and shows the corresponding logarithm of the vapor pressure.

nitude lower than the effective partial pressure within the samples.

In Figure 3 we plot constant strength lines in Temperature -- log $(P/\dot{S})$ space. The water vapor pressure is plotted such that the appropriate strength corresponds to a stress rate of 10^{-7} MPa/s (right hand ordinate). If we assume a modulus of 100 GPa for the rock, the corresponding strain rate is 10^{-12} /s, a geologically significant strain rate. At a depth of 10 km in the crust we expect temperatures to be on the order of 250 to 300° C (< 600 K). If the strain accumulation is uniform at 10^{-12}/s and the rocks are saturated with water vapor, we find negligible strength. Not until either the vapor pressure is decreased or the strain rate increased (by 6 orders of magnitude) do the rocks reach a strength of 10 MPa (100 bars), a typical value for a stress drop at that depth as inferred by seismologists. A further increase in strain rate of 2 orders of magnitude would result in a strength of 100 MPA (1 kbar).

To the extent that this mechanism is important in the failure of rock associated with earthquakes we would expect the following: (1) no earthquakes below a depth where rock temperatures exceed 600 K unless the rocks are extremely dry or are deformed at very high rates; (2) rock failure occurring at high temperature (greater depth) should be associated with small stress drops. As the rock fractures and the fault trace propagates upward, the stress drops should increase; (4) the highest stress drops should occur on asperities that are loaded very quickly during the fault propagation. Note that according to Figure 3 the shallower the depth (i.e. the lower the temperature), the larger is the increase in stress rate needed to increase the failure strength by a given factor.

Several points should be made as to the quantitative limits of the applicability of Figure 3. At the high strength limit, failure may occur by a different mechanism. For example a strength of 1 GPa (100 kbars) may never be realized at T = 300 K and log $[P^n/\dot{S}] = -2$ without the application of confining pressure. The effect of confining pressure was not considered in this paper. It would in general, through its influence upon the crack morphology, increase the strength but should not greatly affect the temperature dependence of the strength since activation energies are typically slowly varying functions of confining pressure (Nachtrieb, et al. 1959).

At high temperatures and very slow stress rates, the crack tips may become more blunt, thereby reducing the crack tip radii and increasing the stress necessary to maintain a constant crack tip velocity. The effect of the blunting would be to increase the strength at high temperatures. Single crack propagation and rock strength tests over a large range in temperature and stress rate are needed to improve the extrapolation to geological time scales. Because of the difficulties in drying more conventional rock specimens, the moisture effect studies must probably be confined to single crack propagation experiments in thin plates on glass and rock.

Acknowledgments: This work was partially supported by the U.S.G.S. through contract 14-08-0001-18307 and NASA through grant number NSG7584.

REFERENCES

Anderson, O.L., Grew, P.C., 1977, Stress corrosion theory of crack propagation with applications to geophysics, *Rev. Geophys. Space Phys.*, **15,** 77-104.

Atkinson, B.K., 1979, A fracture mechanics study of subcritical tensile cracking of quartz in wet environments, *Pure Appl. Geophys.*, in press.

Brodsky, N., Spetzler, H., 1979, Time dependent deformation of a basalt at low differential stress, to be published in *Proceedings of the Lunar and Plantetary Sci. Conf.* **10th.**

Hillig, W.B., Charles, R.J., 1965, in High Strength Materials, edited by V.F. Zackey, John Wiley & Sons, Inc., New York.

Kranz, R.L., 1979, Crack-crack and crack-pore interactions in stressed granite, *Int. J. Rock Mech.*, **16,** 37-47.

Martin, R.J. III, 1972, Time-dependent crack growth in quartz and its application to creep in rocks, *J. Geophys. Res.*, **77,** 1406-19.

Martin, R.J. III, Durham, W.B., 1975, Mechanisms of crack growth in quartz, *J. Geophys. Res.*, **80,** 4837-4844.

Mizutani, H., Spetzler, H., Getting, I., Martin, R.J. III and Soga, N., 1977, The effect of out gassing upon the closure of cracks and the strength of lunar analogues, *Proc. Lunar Sci. Conf.* **8th,** 1235-1248.

Nachtrieb, N.H., Resing, H.A. and Rice, S.A., 1959, Effect of Pressure on self-diffusion in lead, *J. of Chem. Physics*, **31,** 135-38.

Rudnicki, J.W., 1979, The stabilization of slip on a narrow weakening fault zone by coupled deformation - pore fluid diffusion, *Bull. Seismol. Soc. Am.*, **69,** 1011-26.

Scholz, C.H., 1972, Static fatigue of quartz, *J. Geophys. Res.*, **77,** 2104-14.

Scholz, C.H., Koczynski, T.A., 1979, Dilatancy anisotropy and the response of rock to large cyclic loads, *J. Geophys. Res.*, **84,** 5525-34.

Siegried, R., Simmons, G., 1978, Characterization of oriented cracks with differential strain analysis, *J. Geophys. Res.*, **83,** 1269-78.

Soga, N., Okamoto, T., Hanada, T., and Kunungi, M., 1979, Chemical reaction between water vapor and stressed glass, *J. Am. Ceram. Soc.*, **62,** 309-310.

Swanson, P., 1979, Master's thesis, Univ. of Colorado, Stress corrosion cracking rocks: examined by the double- torsion technique.

Swanson, P., Spetzler, H., 1979, Stress corrosion of single cracks in flat plates of rock, *EOS, Transactions, Am. Geophys. Union*, 60-380.

Tapponnier, P., Brace, W.F., 1976, Development of stress- induced microcracks in westerly granite, *Rock Mech. Min. Sci. and Geomech. Abstr.*, **13,** 103-12.

Wiederhorn, S.M., 1967, Influence of water vapor on crack propagation in soda-lime glass, *J. Am. Ceram. Soc.*, **50,** 407-14.

Widerhorn, S.M., 1969, Fracture of ceramics, mechanical and thermal properties of ceramics, *Nat. Bur. Stand. Publ.*, **303,** 217-41.

MARTENSITIC OLIVINE-SPINEL TRANSFORMATION AND PLASTICITY OF THE MANTLE TRANSITION ZONE

J.P. Poirier

Institut de Physique du Globe, Université Paris VI
4, Place Jussieu, 75230 Paris Cedex 05.

Abstract. From a structural analysis of the transition from the olivine lattice to the spinel lattice and by analogy with what is well known for the similar transitions in cobalt (hcp to fcc) and austenitic strainless steels (formation of ε martensite), we propose that the olivine spinel transition is a martensitic diffusionless shear transformations (as opposed to a nucleation and growth transformation controlled by oxygen diffusion). It therefore must be stress sensitive and occur close to equilibrium in the transition zone of the mantle. It is effected by motion of partial dislocations, corresponding to splitting of the dislocations of the (100) [001] glide system in olivine and {111} <110> in spinel, and trailing wide stacking faults in the parent lattice.

The shear modulus corresponding to these slip systems should decrease considerably near the transition.

Transformation plasticity in the transition zone is predicted to occur by abundant and easy slip on the (100) [001] system in olivine and {111} <110> in spinel.

1. Introduction

When polycristalline solids exhibiting a phase transition are driven under stress through the transition temperature or pressure, the creep rate is enhanced during the time the transition takes place (see Poirier, 1976). This is known as transformation superplasticity or better as transformation-induced plasticity. Greenwood and Johnson (1965) have accounted for this effect in the following way : the volume change in grains undergoing the transition creates internal stresses in the neighboring material ; when these stresses reach the elastic limit of the material, they can be relieved by plastic flow, directed by the applied stress. This mechanical explanation applies to all types of phase transition provided there is a volume change, and accounts for the Newtonian behavior of the material when processed back and forth through the transition. Gordon (1971) and Sammis and Dein (1974) have suggested that this process may take place in the Earth's mantle when olivine transforms to spinel. We will try here to take a physical (as opposed to mechanical) view of the problem and assess the importance of transformation-induced plasticity in the transition zone of Earth's mantle where material in the ascending or descending limbs of convection cells undergoes phase transitions over a depth range of nearly 300 km.

2. The olivine-spinel phase transition

We will consider here the olivine-γspinel transition, for although the β spinel phase is certainly a major phase in the transition zone, it is γ spinel which is first formed from olivine and later transforms to β spinel, as can be seen by simple inspection of the phase diagram (Ringwood and Major, 1970 ; Akimoto et al., 1976). It is usually accepted that the olivine-spinel transition operates by nucleation and growth and that it is controlled by the diffusion of oxygen ions (Sung and Burns, 1976 a, b) ; we will argue here that it is a martensitic diffusionless transformation which operates by dislocations.

The lattice of oxygen ions in olivine is hexagonal close packed and it is face centered cubic in γ spinel. Clearly the controlling stage in the phase transition is the rearrangement of the lattice of bulky oxygen ions. Now this can be done easily by sweeping a partial dislocation across the basal plane of the hcp lattice thus introducting a layer of fcc stacking fault ; indeed the necessary partial dislocation can be found by splitting of the dislocations responsible for the (100) [001] slip system of olivine. The (100) plane is the basal plane of the hcp lattice and the [001] dislocation can be split according to the reaction :

$$[001] \rightarrow \frac{1}{12}[013] + \frac{1}{12}[0\bar{1}3] + \frac{1}{12}[013] + \frac{1}{12}[0\bar{1}3]$$

The first partial introduces a stacking fault in the oxygen lattice, and the Mg^{++} cations find themselves in the tetrahedal sites whereas some of the Si^{4+} cations find themselves in octahedral sites. The cations therefore must rearrange themselves by jumping into neighboring proper sites which are γ spinel sites (Fig. 1). The movement of the first partial corresponding to (100) [001] glide thus brings about a layer of stacking fault with spinel structure. The process of rearrangement of cations by one atomic distance jumps is known as synchroshear ; it was first proposed by Kronberg (1957) for slip in sapphire and applied by Hornstra (1960) to slip

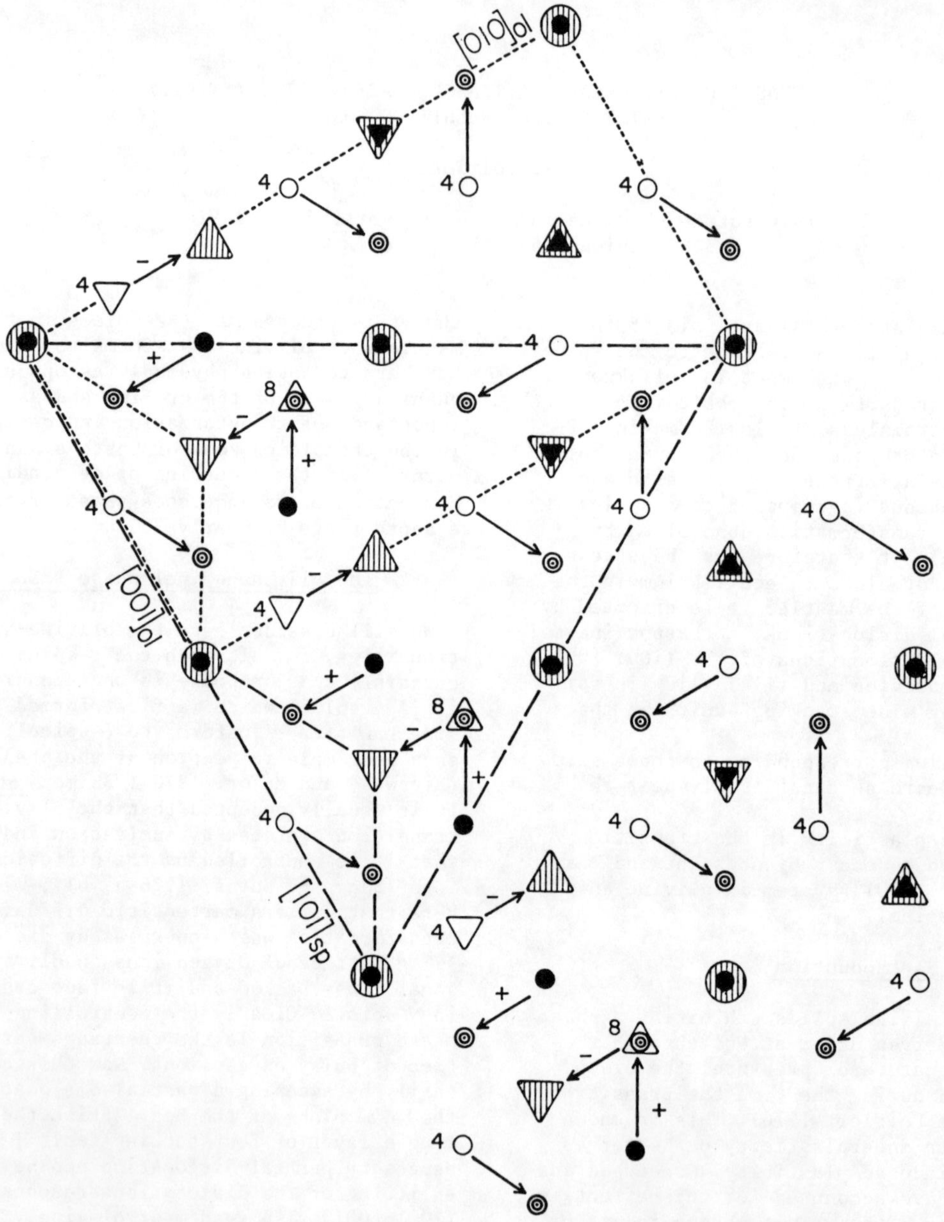

Fig. 1. Scheme of the olivine-spinel transformation by dislocations. The plane of the figure is the close-packed plane : (100) for olivine, (111) for spinel. Closed triangles : Positions of Si cations in olivine at level 0. Open triangles : Positions of Si cations in olivine at level $\frac{1}{2}$ (100). Closed circles : Positions of Mg cations in olivine at level 0. Open circles : Positions of Mg cations in olivine at level $\frac{1}{2}$ (100). Hatched triangles : Positions of Si cations in the spinel mixed layer. Hatched triangles : Positions of Mg cations in the spinel mixed layer. Double circles: Positions of Mg cations in the spinel kagome layer. Arrows indicate the sense and length of the shuffles of ions. + or - signs indicate that the ion moves towards the upper or lower level. Numerals (4 or 8) near the symbols indicate that the site has become respectively tetrahedral or octahedral.

in the spinel lattice. Synchroshear involves a coordinated shuffle of the cations and has nothing to do with diffusion. The second partial erases the stacking fault in the oxygen lattice by restoring a hcp structure, the cations shuffle back to their proper olivine sites by a synchroshear inverse of the previous one, but there remains a stacking fault in the order of the cations (i.e. an antiphase boundary) ; the third partial once again produces a spinel layer and the fourth one restores the original olivine lattice. The width of the stacking fault ribbons between the successive partials depends on the stacking fault energies. As the transition is approached, the spinel structure becomes the stable one and the first and third stacking fault ribbons tend to become very wide as a result of the stacking fault energy being lowered. The olivine crystal can be invaded by these stacking faults, i.e. by layers of spinel, thus bringing about the transformation.

The inverse process from spinel to olivine takes place by dissociation of $\langle 110 \rangle$ dislocations on $\{111\}$ glide planes of spinel according to the reaction :

$$\frac{1}{2}[10\bar{1}] \rightarrow \frac{1}{12}[11\bar{2}] + \frac{1}{12}[\bar{2}11] + \frac{1}{12}[11\bar{2}] + \frac{1}{12}[\bar{2}11]$$

The stacking fault has then the usual displacement vector $\frac{1}{3}[111]$. Indeed, the idea that the spinel-olivine transition could take place by such a mechanism had already been proposed by Hornstra (1960) but had apparently gone unnoticed.

Such a diffusionless shear transformation is usually termed martensitic. It is well documented in the case of the metal cobalt which undergoes a phase transition form hcp to fcc at about 400°C (Christian, 1951 ; Bibring and Sebilleau, 1955). The phase transformation has been directly observed by Transmission Electron Microscopy and it has been clearly proved that it proceeds by invasion of the parent crystal by sheets of stacking fault trailing behind partial dislocations on the close-packed planes (Votava, 1960 ; Bollman, 1961). The stacking fault energy has been measured and found to decrease considerably as the transition temperature was approached (Ericsson, 1966). Similar observations in TEM were made for the formation of hcp ε martensite in fcc austenitic strainless steels (e.g. 18 % Cr, 8 % Ni) (Reed, 1962 ; Venables, 1962 ; Brooks et al, 1979). Brooks et al, 1979 have shown that the partial dislocations effecting the transition were nucleated at the grain boundaries : moreover they were able to determine by contrast analysis that even single stacking faults had a supplementary displacement corresponding to the contraction of about 1 %, normal to the close packed plane accompanying the formation of the ε phase. Thus, the layer of hcp material produced by the passage of the partial dislocation in the fcc austenite is in all respects a layer of ε phase. This observation is especially relevant in the case of the olivine spinel transition since here too, in going for olivine to spinel there is a contraction of about 1 % normal to the close packed planes.

So there is ample evidence that the transition between hcp and fcc lattices occur by motion of partial dislocations on the close packed planes : according to the unwritten rule that a transition which can be effected by dislocations is effected by dislocations (and not by destruction and reconstruction of the lattice by diffusion) we think that the transition olivine-spinel occurs in this fashion for the oxygen lattice and is automatically accompanied by the proper synchroshear, which is anyway necessary in the case of the (100) [100] slip of olivine. This is the easiest mechanism available and we do not see any good reason to resort to a nucleation and growth mechanism controlled by diffusion (least of all by diffusion of oxygen). The only diffusion one might need would be a possible diffusion of Fe^{++} cations to obtain the correct equilibrium ratio Fe^{++}/Mg^{++}, but even this is not really necessary as it is possible to form spinel in conditions of chemical non-equilibrium.

A number of consequences follow :

i) the olivine-spinel transition must be assisted by shear stresses creating a driving force for the motion of the leading partial dislocations.

ii) As in the case of cobalt (Bibring and Sébilleau, 1955) we may expect a wide hysteresis cycle : the olivine-spinel transition must be easier than the reverse one since olivine having anisotropic linear compressibilities and thermal expansion coefficients, polycrystals will develop internal stresses when P and T increase, thus assisting the transformation. Such will not be the case when isotropic spinel is decomposed.

iii) It is possible (and even probable) that the shear transition, effected by dislocations, becomes possible because of a lattice instability corresponding to a phonon "soft mode" of low frequency (Clapp, 1973).

The relevant elastic constants are then C_{66} and C_{55} corresponding to shear along the (100) plane. These constants (or a linear combination of C_{66}, C_{55} and higher order constants) should decrease as the transition is approached. Nothing is known of the elastic constants of olivine at high pressures but we can notice that C_{66} and C_{55} have the smallest dC/dP increase with pressure about room pressure (Kumazawa and Anderson, 1969).

3. Transformation plasticity

If the olivine-spinel transition is a diffusionless shear transformation effected by dislocations, then there must be a strong coupling between phase transition and plastic deformation, as is indeed the case for cobalt (de Lamotte and Alstetter, 1969 ; Holt and Teghtsoonian, 1972). The partial dislocations responsible for the transformation would be the one normally constituting the dissociated dislocations responsible for (100) [001] slip in olivine and (111) [110] slip in spinel. Although the existence of (111) [110] slip in $(Mg, Fe)_2 SiO_4$ spinel has not yet been proved, there are at least indications that it may be active, and dislocations with [110] Burgers vector have been observed (Madon and Poirier, 1980). As for (100) [001] slip in olivine it is well documented as a low temperature ($T < 1000°C$) slip system (Nicolas and Poirier, 1976) whereas (010) [100] is more active at high temperatures ; it would therefore seem that (100) [001] slip could not be important at the high temperatures expected in the transition zone. However we must note that all experiments have been done at pressures lower than 20 kbars and that (100) [001] slip may be active at high temperatures and high pressures such as 100 kbars. Indeed this is the system we should expect to find in abundance as the transition is approached since it provides the proper shear for an energetically favorable transition. As the Clapeyron slope is positive, the phase boundary can be reached from a point in the olivine domain by lowering T or increasing P and we can predict that the (100) [001] systems should be dominant at the temperatures and pressures of the deeper peridotite mantle. We should also note that screw dislocations with [001] Burgers vector are the only ones to be observed in untransformed olivine in the shocked meteorite Tenham where spinel was found (Madon, personal communication).

The shear modulus relevant to (100) [001] slip is C_{55} which we expect to decrease considerably near the transition. This would account for an easy multiplication of [001] dislocations since the critical stress σ_C for the activation of a source of length l,

$$\sigma_C \simeq \frac{C_{55}b}{l}$$ would decrease.

We therefore expect the appearance of transformation plasticity even before the transition and we suggest that it takes place by abundant and easy glide on the (100) [001] slip system of olivine in grains which are about to transform into spinel.

4. Conclusions

From a structural analysis of the olivine-spinel transition and by analogy with similar transitions in metals we suggest that :

1. The olivine-spinel transition is a shear diffusionless transition (martensitic), and as such is stress-sensitive.
2. It is effected by notions of partial dislocations trailing sheets of stacking fault in the oxygen lattice, the cations being rearranged by synchroshear.
3. The transition in the sense olivine-spinel occurs close to the equilibrium.
4. The elastic constants C_{66} and C_{65} of olivine should decrease considerably near the transition.
5. Transformation plasticity in the transition zone occurs by abundant and easy slip on the (100) [001] system in olivine and {111} <110> system in spinel.

Contribution IPG NS 381.

References

Akimoto, S., Y. Matsui, and Y. Syono, High pressure chemistry of orthosilicates and the formation of the mantle transition zone, The Physics and Chemistry of Minerals and Rocks, edited by R.G.J. Strens, 327-363, Wiley, 1976.

Bibring, H., and F. Sebilleau, Structure et transformation allotropique du cobalt, Rev. Metallurgie, 52, 569-578, 1955.

Bollmann, W., On the phase transformation of cobalt, Acta Metall., 9, 972-975, 1961.

Brooks, J.W., M.H. Loretto, and R.E. Smallman, In situ observation of the formation of martensite in stainless steels, Acta Metall., 27 1829-1838, 1079.

Brooks, J.W., M.H. Loretto, and R.E. Smallman, Direct observation of martensite nuclei in stainless steel, Acta Metall., 27, 1839-1847.

Christian, J.W., A theory of the transformation in pure cobalt, Proc. Roy. Soc., A. 206, 51-64, 1951.

Clapp, P.C., A localized soft mode theory for martensitic transformation, Phys. Stat. Sol. (b) 57, 561-569, 1973.

Demarest, H.M., R. Ota and O.L. Anderson, Prediction of high pressure phase transition by elastic constant data. High Pressure research, edited by M. Manghnani and S. Akimoto, Acad. Press, 1977.

Ericsson, T., The temperature and contractions dependence of the stacking fault energy in the Co-Ni system, Acta Metall., 14, 853-865, 1966.

Fujita, H., and S. Ueda, Stacking faults and Fcc (γ) $\rightarrow$ hcp (ε) transformation in 18/8 type stainless steel, Acta Metall., 20, 759-767, 1972.

Gordon, R.B., Observation of crystal plasticity under high pressure with applications to the Earth's mantle, J. Geophys. Res., 76, 1248-1254, 1971.

Greenwood, G.W., and R.H. Johnson, The deformation of metals under small stresses during phase transformations, Proc. Roy. Soc. London, A. 283, 403-422, 1965.

Holt, R.T., and E. Teghtsoonian, The tensile deformation of cobalt single crystals in the Fcc phase, Met. Trans., 3, 1621-1626, 1972.

Hornstra, J., Dislocations, stacking faults and twins in the spinel structure, J. Phys. Chem. Solids 15, 311-323, 1961.

Kelly, P.M., The martensitic transformation in steels with low stacking fault energy. Acta Metall. 13, 635-646, 1965.

Kronberg, M.L., Plastic deformation of single crystals of sapphire ; Basal slip and twinning, Acta Metall., 5, 507-524, 1957.
Kumazawa, M., and O.L. Anderson, Elastic moduli, pressure derivatives and temperatures derivatives of single crystal olivine and single crystal forsterite, J. Geophys. Res., 74, 5961-5972, 1969.
De Lamotte, E., and C. Alstetter, Transformation strain in stressed cobalt-nickel single crystals, Trans AIME, 245, 651-659, 1969.
Madon, M., and J.P. Poirier, Dislocations in spinel and garnet high pressure polymorphs of olivine and pyroxene. Implications for mantle rheology, Science, 207, 4426, 1980.
Nicolas, A., and J.P. Poirier, Crystalline plasticity and solid state flow in metamorphic rocks, Wiley, Interscience, 444, 1976.
Poirier, J.P., Plasticité à haute température des solides cristallins, Eyrolles, 320, 1976.
Reed, R.P., The spontaneous martensitic transformations in 18 % Cr 8 % Ni steels. Acta Metall., 10, 865-877.
Ringwood, A.E., and A. Major, The system Mg_2SiO_4-Fe_2SiO_4 at high pressures and temperatures, Phys. Earth Planet. Interiors, 3, 89-108.
Sammis, C.G., and J.L. Dein, On the possibility of transformation superplasticity in the Earth's mantle. J. Geophys. Res., 79, 2961-2965.
Sung, C.M., and R.G. Burns, Kinetics of the olivine-spinel transition : implications to deep focus earthquake genesis, Earth and Planet. Sci. Lett., 32, 165-170, 1976.
Venables, J.A., The martensite transformation in stainless steels, Philos. Mag., 7, 35-44.
Votava, E., Electron microscopic investigation of the phase transformation of thin cobalt samples. Acta metall., 8, 901-904.

ULTRASONIC VELOCITY AND ATTENUATION IN BASALT MELT

Murli H. Manghnani, Chandra S. Rai and Keith W. Katahara

Hawaii Institute of Geophysics, University of Hawaii, Honolulu, Hawaii 96822

Gary R. Olhoeft

U.S. Geological Survey, Denver, Colorado 80225

Abstract. A continuous-wave interferometric method has been developed for measurement of ultrasonic velocity and attenuation in rock melts. This method provides an alternative to the pulse transmission techniques used in previous studies. The feasibility of the method is demonstrated by measurements on water at room temperature and by exploratory experiments on a basalt melt at 1300-1400°C. Advantages, sources of error, and possible improvements of the method are discussed.

Introduction

Ultrasonic wave velocity, V, and attenuation, Q^{-1}, in molten and partially molten rocks are parameters of significant interest in geophysics. They are relevant to seismological exploration of volcanically active areas such as mid-ocean ridges, island arcs, and possible sources of geothermal energy. Experimental data on velocity and attenuation for partial melts have an important bearing on discussions of the low-velocity, high-attenuation zones in the upper mantle where partial melt may exist. Such measurements would also help to determine the structure of silicate melts and to clarify the physical processes involved in such related melt properties as viscosity.

Perhaps because of the experimental difficulties involved, few measurements of V and Q have been reported for partially or fully molten rocks, in spite of the importance of such measurements. There is considerable literature on some simple alkali silicate and borate systems [see, for example, Bloom and Bockris, 1957; Kumazawa *et al*., 1964; Simmons and Macedo, 1968; Tauke *et al*., 1968; Laberge *et al*., 1973]. For silicate melts of geophysical importance, however, the only work that has been reported is that of Murase and coworkers [see Murase and Suzuki, 1966; Murase and McBirney, 1973; Murase *et al*., 1977].

Previous laboratory studies of melt velocity and attenuation have primarily used pulse transmission techniques in which velocities are obtained by time-of-flight measurements. In this note we describe a simple, continuous-wave interferometric method that is a promising alternative to pulse transmission methods, and we present preliminary measurements on a basalt melt which demonstrate the feasibility of the method.

Method

A schematic diagram of the apparatus is shown in Figure 1. The physical arrangement is similar to that of previous work [Murase and McBirney, 1973] in that buffer rods are used to conduct the ultrasonic waves into and out of the melt contained in a crucible attached to the lower rod. Ultrasonic waves are generated and received by water-cooled piezoelectric transducers attached to the ends of the rods.

Our method is based on the fact that the melt layer is in a quasi-resonant state when its thickness, ℓ, is a multiple of half the wavelength, λ. The apparatus is designed to observe the resonances and antiresonances as either the melt thickness or wavelength is changed. Thus in Figure 1 an oscillator produces a continuous sine wave that is applied to the upper transducer. The frequency is read by a digital counter. After passing through the rods and the melt, the ultrasonic wave is picked up by the lower transducer and the signal is amplified, rectified, and applied to the Y-axis of an X-Y recorder. The melt layer thickness is controlled by raising and lowering the upper rod by a motor. The rod position is determined by a linear displacement transducer (LDT) calibrated by a dial gauge. The DC voltage output of the LDT is applied to the X-axis of the recorder so that a plot of transmitted amplitude against melt layer thickness is obtained as the rod is moved up or down. Alternatively, the frequency, and thus the wavelength, can be varied at fixed layer thickness. The latter method would be used at temperatures at which the sample is solid, partially molten, or very viscous.

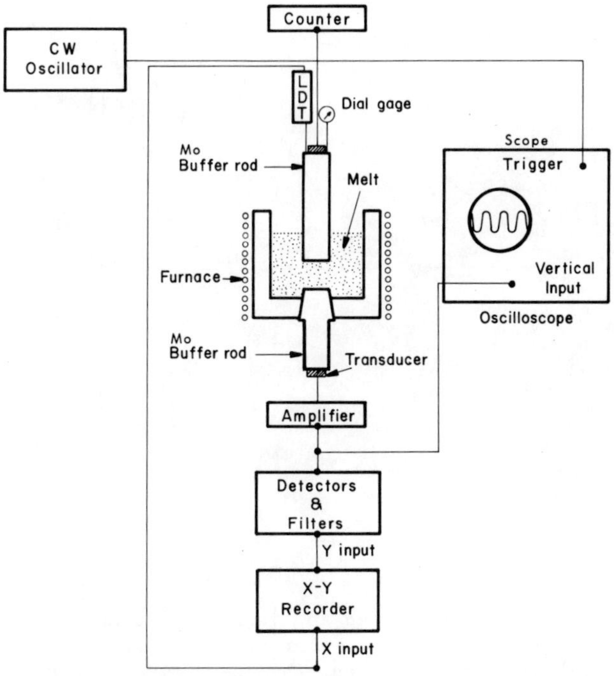

Fig. 1. Schematic diagram of apparatus for the continuous-wave technique of measuring velocity and attenuation.

The resulting plots show amplitude peaks near the resonant conditions $\ell = N\lambda/2$. For fixed frequency, the pattern of alternating maxima and minima, of course, decays with increasing length because of attenuation in the melt. When the frequency rather than the length is swept, plots must be taken at several lengths in order to evaluate the effects of attenuation.

The usefulness of this method depends in part on the contrast in acoustic impedance (density times velocity) between the melt and the rods. As shown below, a large impedance contrast gives rise to a large difference between amplitude maxima and minima, and thus makes measurements easier. In the present case, the buffer rods were single crystals of molybdenum, 1.27 cm in diameter by 20 cm long, chosen for high impedance and low attenuation at high temperatures. Molybdenum buffer rods are also less reactive with the silicate melt, under reducing conditions, as compared to quartz rods used by Murase and McBirney [1973].

The velocity, V, and the quality factor, Q, can be extracted from the data as follows. Assume that the wavelength is much smaller than the rod diameter, and that multiple reflections within the rods can be neglected. (This latter condition is not well satisfied in the work described here, but it can be easily fulfilled by a modification to the experiment as will be discussed below.) We make the additional reasonable assumption that the attenuation in the molybdenum rods is negligible compared with the attenuation in the melt.

We then consider the one-dimensional problem shown in Figure 2 in which a continuous plane wave is incident on the melt layer from the left, and we solve for the amplitude of the transmitted wave using conventional elasticity theory.

The details of the derivation are given elsewhere [Katahara et al., unpublished]. The result for the transmitted amplitude, A_t, relative to the incident amplitude, A_i, is

$$\left|\frac{A_t}{A_i}\right| = 4\Big/\left[2(4 + W)\cosh\frac{\beta\ell}{Q} + 8U\sinh\frac{\beta\ell}{Q} + 2(4 - W)\cos 2\beta\ell - 8X\sin 2\beta\ell\right]^{\frac{1}{2}} \quad (1)$$

where

$$U = Z_r + \frac{1}{Z_r\left(1 + \frac{1}{4Q^2}\right)} \quad (2)$$

$$X = \left[Z_r - \frac{1}{Z_r\left(1 + \frac{1}{4Q^2}\right)}\right] \Big/ 2Q \quad (3)$$

$$W = U^2 + X^2 \quad (4)$$

$$Z_r = \frac{\text{ROD IMPEDANCE}}{\text{MELT IMPEDANCE}} \quad (5)$$

$$\beta = \text{MELT WAVENUMBER} = \frac{2\pi f}{V} \quad (6)$$

where ℓ is the melt thickness, V is the velocity, and f is the frequency.

The quantity $|A_t/A_i|$ in equation (1) is what is measured in our experiment. Since it is a rather complicated expression, consider first the sim-

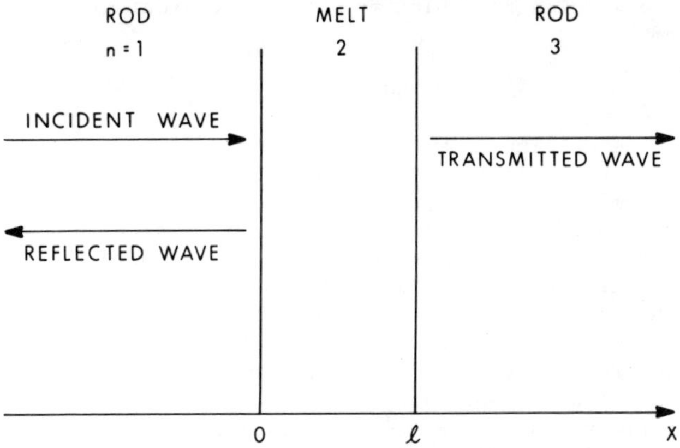

Fig. 2. Diagram showing idealized rod-melt system in which an infinite train of monochromatic plane wave is normally incident on the melt which is assumed to be infinite in the direction perpendicular to the x-axis.

plified case in which there is no attenuation and $Q^{-1} = 0$. Equation (1) reduces to

$$\frac{A_t}{A_i} = \frac{1}{\left[1 + \left(\frac{Z_r^2 - 1}{2Z_r}\right)^2 \sin^2 \beta\ell\right]^{\frac{1}{2}}} \quad (7)$$

This is a periodic function with maxima

$$\left|\frac{A_t}{A_i}\right| = 1 \quad (8)$$

for integral N, and minima

$$\left|\frac{A_t}{A_i}\right| = \frac{2Z_r}{Z_r^2 + 1} \quad (9)$$

Thus there is perfect transmission at the maxima when the melt layer thickness is $N\lambda/2$. This situation is completely analogous to the case in optics involving multiple reflections in a thin layer [see Jenkins and White, 1957, chapter 14, in which their equation 14j is analogous to our equation (7)]. At each reflection the wave is partially reflected, and the layer becomes transparent when all the transmitted components are in phase. From (9), the minima $\ell = (N + 1/2)(\lambda/2)$ have amplitudes that depend on the impedance ratio Z_r. The minima become deeper and the transmission peaks become sharper as the impedance contrast increases. Since deeper minima and sharper peaks are easier to observe experimentally, it follows that one should use high impedance materials, such as molybdenum, for the buffer rods. Note that velocity measurements can be carried out with length as small as one wavelength, which may be of the order of 1 mm.

Results and Discussion

The continuous wave technique has been used in several exploratory measurements. Figure 3 shows a comparison of experimental and theoretical curves of amplitude plotted against length for water at room temperature at 3.6 MHz. Note that the transmitted amplitude peaks are very sharp. This is a consequence of the high impedance ratio, $Z_r = 44$, between molybdenum and water. The theoretical peaks are sharper than the experimental peaks, probably because of multiple reflections in the buffer rods, nonparallelism of the rods and distortion of the wave fronts due to diffraction and inhomogeneities in the rods and transducers. For high-Q samples such as water, poor rod alignment is particularly troublesome because the waves undergo many multiple reflections in the liquid layer and the wavefronts rotate farther out of the transducer plane at each reflection.

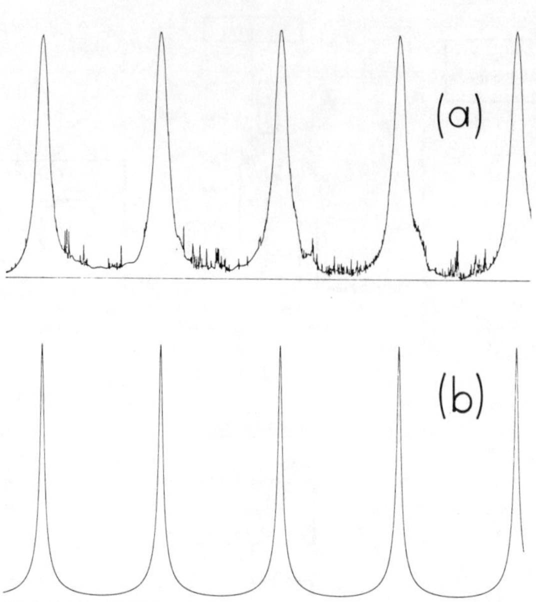

Fig. 3. Transmitted amplitude of 3.6 MHz versus layer thickness for water at room temperature. (a) experimental data; (b) theoretical curve based on equation (1).

We have obtained experimental data on room-temperature water extending out to 50 transmission peaks at 5.66 MHz. There was a barely-observable decay in the amplitude-length records, which corresponds to Q^{-1} of about 0.0002. The American Institute of Physics Handbook value for water is 0.00007, so the present apparatus is capable of measuring Q^{-1} values to within 0.0001 or 0.0002 at room temperature.

Figure 4 shows some digitized data from an exploratory experiment on an alkalic basalt melt,

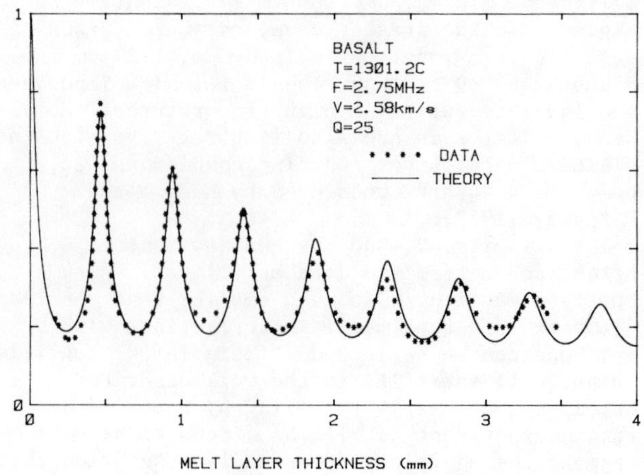

Fig. 4. Preliminary experimental data on basalt melt at 2.75 MHz.

and the theoretical curves resulting from the fitting procedure. The fit was obtained by a least-square procedure with five adjustable parameters: Q, V, the origins of the amplitude and thickness scales, and a scaling factor for the amplitude scale. Zero melt thickness is impossible to achieve because of the viscosity of the melt, so in any work on melts the thickness origin and the zero-thickness amplitude need to be floating parameters. (The zero-amplitude point was not measured experimentally because it was not thought to be significant at the time the data were taken.) The velocity values are not in any case sensitive to the fitting procedure and can be obtained to a good approximation simply from the peak positions. Measurements taken over the range of temperatures 1300-1400°C and at 2.7-3.7 MHz are listed in Table 1. The velocities, which are accurate to within 2-3%, show no significant temperature or frequency dependence over this range. This result is in agreement with those of Murase and McBirney, who found velocities to be independent of temperature above 1200°C for a number of melts. The reported Q values should be treated with caution as the present data are not completely satisfactory because of multiple reflections in the molybdenum rods, poor alignment of the rods, nonlinearity in the rectifier circuit, limited recorder frequency response, reflections from the sides of the crucible, and jerkiness in the motor that positions the movable rod. These factors, which are very difficult to analyze, will have significant effect on the Q values but not on the velocity values. Errors in Q on the order of 50% are easily conceivable. However, most of these errors can be significantly diminished by straightforward improvements to the apparatus. Note that it is possible in principle to determine the melt density by letting it be an adjustable parameter in the fitting procedure. This cannot be done in practice because the fit is not very sensitive to the density.

This method can also be used in a pulsed mode in which the pulses are long enough to achieve quasi-steady-state conditions (a few tens of microseconds are sufficient for melts at MHz frequencies). The amplitude near the end of the transmitted pulse will then correspond to equation (1). This arrangement is electronically more complex but is more flexible in the frequencies that can be used, and it would be more useful in situations in which the melt layer is not variable. It would also eliminate the problem of multiple reflections in the buffer rod.

TABLE 1. Longitudinal Velocity and Quality factor of an Alkalic Olivine Basalt as Function of Temperature and Frequency

T, °C	f, MHz	Q	V_p, km/sec.
1295.0	3.66	15.5	2.62
1297.5	2.53	19.4	2.53
1301.1	3.68	17.5	2.56
1301.2	2.75	30.0	2.58
1301.4	3.66	15.4	2.56
1340.6	3.67	11.2	2.69
1364.4	3.67	36.8	2.66
1394.8	2.74	14.6	2.61
1395.9	3.66	21.3	2.66

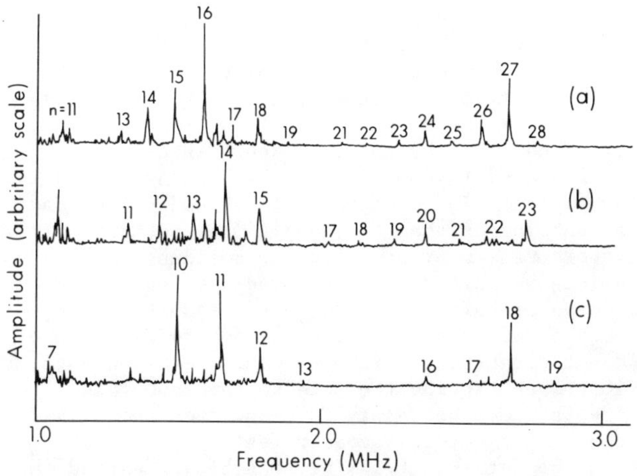

Fig. 5. Transmitted amplitude versus frequency for water layers with thickness ℓ equal to (a) 5.08 mm, (b) 6.35 mm and (c) 7.62 mm. Velocity can be calculated from $V = 2n\ell f$; values of n for observable peaks are shown in the figure.

Recall that the phase, $\beta\ell$, in equation (1) can be varied by changing the frequency as well as the length. Measurements on water, at room temperature, in which the length was fixed and the frequency was swept, are shown in Figure 5. Transmitted-amplitude peaks are similar to those in Figure 3 except that the pattern is modulated by the response of the piezoelectric transducers, with passbands at the transducer resonance and harmonics. This technique should be especially useful in velocity measurements below the solidus. To extract Q values, measurements below the liquidus can be repeated at different lengths.

Summary and Conclusions

We have analyzed and demonstrated the feasibility of a continuous-wave technique for measuring velocity and attenuation in melts. The equipment required is simple and readily available. As opposed to pulse transmission methods requiring high voltage pulses, we obtain satisfactory results with ordinary rf oscillators generating 10-20 volt sine waves. The velocity measurements in particular are easy to carry out and can be made at least as accurately as with pulse methods. Phase velocities rather than group velocities are measured; this factor may be important if dispersion is significant.

This method, of course, also has some drawbacks.

(a) The sample density and the rod impedance need to be known or estimated. The velocity is virtually independent of these quantities, but the Q value will be affected slightly if they are in error. (b) The attenuation of the rod is neglected. If the rod Q is known it can be incorporated in equation (4) by letting k_1 and Z_1 be complex. The Q of Mo or Ta should in any case be much larger than the sample Q except for temperatures near or below the solidus. (c) Various undesired resonance modes in the transducer-buffer rod-crucible assembly may be excited and interfere with the desired signal. (d) At least four parameters must be allowed to float when fitting the data to theory. This affects the uncertainty in Q but does not matter much in the velocity measurement.

In the future, it may be possible to extend the use of this method to shear waves also. This method can be used for relatively small melt layer thicknesses and hence may not require very high gain amplification. Furthermore, in the continuous-wave mode, high signal-to-noise ratios can be easily achieved with a lock-in amplifier without the sophisticated electronics used by Simmons and Macedo [1968] to adapt lock-in amplification to pulse measurements.

Acknowledgments. This work was supported by U.S. Geological Survey Extramural Geothermal Program grant 14008-0001-G-576, and by National Science Foundation grant EAR 78-23838, and by a grant from the Hawaii Natural Energy Institute. We thank J. Balogh for designing and constructing the apparatus and for help in taking the data. We are grateful to Dr. W. Haller, U.S. National Bureau of Standards, for lending us the molybdenum rods. Hawaii Institute of Geophysics Contribution No. 1133.

References

Bloom, H., and J.O'M Bockris, The compressibilities of the silictes: The Li_2O-SiO_2 system, J. Phys. Chem. 61, 515-518, 1957.

Jenkins, F.A., and H.E. White, Fundamentals of Optics, McGraw-Hill, New York, 1957.

Katahara, K.W., C.S. Rai, M.H. Manghnani, and J. Balogh, An interferometric technique for measuring velocity and attenuation in molten and partially molten rocks, submitted to J. Geophys. Res.

Kumazawa, M., H. Furuhashi, and K. Iida, Seismic wave velocity in molten silicates at high temperature, Bull. Volcanological Soc. Japan 9, 17-24, 1964.

Laberge, N.L., V.V. Vasilescu, C.J. Montrose, and P.B. Macedo, Equilibrium compressibilities and density fluctuations in K_2O-SiO_2 glasses, J. Amer. Ceram. Soc. 56, 506-509, 1973.

Murase, T., and T. Suzuki, Ultrasonic velocity of longitudinal waves in molten rocks, J. Faculty Sci. Hokkaido Univ., Japan Ser. VII, 2, 277-285, 1966.

Murase, T., and A.R. McBirney, Properties of some common igneous rocks and their melts at high temperatures, Geol. Soc. Amer. Bull. 84, 3563-3592, 1973.

Murase, T., I. Kushiro, and T. Fujii, Compressional wave velocity in partially molten peridotite, Carnegie Inst. Washington Year Book 76, 416-419, 1977.

Simmons, J.H., and P.B. Macedo, High-temperature shear ultrasonic interferometer, J. Acoust. Soc. Am. 43, 1295-1301, 1968.

Tauke, J., T.A. Litvotiz, and P.B. Macedo, Viscous relaxation and non-Arrhenius behavior in B_2O_3, J. Amer. Ceram. Soc. 51, 158-163, 1968.